Key Questions in Hydrology and Watershed Management

Key Questions in Hydrology and Watershed Management

Leon Bren and Patrick Lane

CABI is a trading name of CAB International

CABI
Nosworthy Way
Wallingford
Oxfordshire OX10 8DE
UK

CABI
WeWork
One Lincoln St
24th Floor
Boston, MA 02111
USA

Tel: +44 (0)1491 832111
Fax: +44 (0)1491 833508
E-mail: info@cabi.org
Website: www.cabi.org

Tel: +1 (617)682-9015
E-mail: cabi-nao@cabi.org

A catalogue record for this book is available from the British Library, London, UK.

Library of Congress Control Number: 2021939722

References to Internet websites (URLs) were accurate at the time of writing.

ISBN-13: 9781789249682 (paperback)
         9781789249699 (ePDF)
         9781789249705 (ePub)

DOI: 10.1079/ 9781789249682.0000

Commissioning Editor: Ward Cooper
Editorial Assistant: Emma McCann
Production Editor: James Bishop

Typeset by SPi, Pondicherry, India
Printed and bound in the UK by Severn, Gloucester

Contents

About the Authors

Leon Bren

Leon is Associate Professor at the University of Melbourne in Victoria, Australia. His basic training was in forestry and forestry science. He earned a PhD in the hydrology of a small mountain stream from the University of Melbourne and then taught and researched at this institution for some decades. Work included paired watershed studies, geometry of stream buffer strips and the hydrology of river red gum forests along Australia's River Murray. He lives in the Victorian provincial city of Ballarat.

Patrick Lane

Patrick is Professor in the School of Ecosystem and Forest Sciences, the University of Melbourne where he has researched and taught for 16 years. Previously he worked in a government research agency and as a consultant. His research interests include the impact of wildfire on hydrologic processes and water supplies, the ecohydrology of eucalypt forests and the effects of climate change on forest hydrologic functioning. He lives in Melbourne and enjoys playing guitar, bike riding and sport.

Preface

All-too-often we take the provision of clean, fresh water and healthy streams from nearby mountains for granted. And yet, health and wealth depend on this vital commodity. Protection of this resource requires excellent knowledge of watershed hydrology and the impacts of land management on this. Related to this is the ability to measure and quantify both the resource and land-use impacts on this. We hope that this little book is a contribution to meeting these necessary goals.

The origins of this book comes from many years of teaching classes in Hydrology and Watershed Management. After some years of teaching, it became apparent that in this, as in other disciplines, an orderly development of skills was necessary for students to progress. We authors have had or produced other texts which lay out much of the information, but always felt that there was a need for questions for neophyte (and practising) watershed managers to 'cut their teeth on'. Hence we were pleased when CABI asked if we would care to write this book of Questions and Answers.

The questions assume some basic quantitative skills and knowledge of hydraulic and groundwater processes. We have stopped short of more complex material. In compiling these questions, we have been struck by the patient and efficient assembling of information by scientists in the last three centuries. To students we would say that nothing beats a patient study of the discipline (and the related discipline of doing the basic arithmetic). In putting the material together, we found that the Internet was a great source of information with many fascinating examples. Our suggestion is to start at any point in the book, work through examples, and as necessary discuss them with your professors.

For lecturers, we hope that the questions and exercises will give ideas for lectures and practical sessions and eases the never-ending task of finding questions for tests and exams. We found that putting the questions together was a thought-provoking task ('that would make a good topic for a class discussion'), so we hope that you can harvest some of this, too. Although we have had the ideal of 'binary' answers to the questions, we are aware that some answers might be 'arguable' or a matter of opinion, and feel that this is good material for discussion, too.

Finally, to students, enjoy your studies and get your assignments in on time. It's a great time of life and we wish that we were doing it all over again.

Leon Bren
Patrick Lane

Acknowledgements

We'd like to acknowledge the contributions of our wives and families during the preparation of this book. Writing a book means long hours sitting at a computer and the family put up with this and patiently stepped around the barricade of hydrology books that, daily, grew around the desks.

In putting together these questions we were humbled by the work of past scientists who have brought together the knowledge of hydrology and watershed management. With occasional exceptions, to keep the text 'clean', we have not acknowledged sources but, for every question, there are one or more scientists who could (or, perhaps, should) be acknowledged. More specifically, Professor Tim Fletcher of The University of Melbourne spruced us up on urban hydrology – thanks Tim. And, finally, thanks to those scientists who, when asked, allowed their illustration to be used in one of the questions of Chapter 5.

Notes for Users

The book is aimed at students who hope to make a career in watershed science. It is assumed that they have some knowledge of physics, chemistry, soils, geology, and perhaps hydraulics. The book is aimed at allowing students to self-test their own knowledge and perhaps to define areas where they need to do a little 'polishing'. We have not included problems with long computation sequences because we feel that they are best done with a professor present and/or using a relevant computer package.

This book includes five sections where each section contains a certain type of questions. These are divided into multiple choice questions, matching questions, fill-in-the-blank questions, true/false questions, and image-based questions. Some of the questions test knowledge but we have also tried to test understanding and judgement skills. Ideally, the questions are aimed at making the student 'think'. In posing the questions we are aware that some of them are complex and could, themselves, be expanded to their own book. Some can be readily answered by recourse to a reference book but it will not be so easy for others. In writing these questions we often felt guilty that we had not cited someone or other's research, but the absence of references makes the text cleaner; we hope that the readers will accept at face value that we could, if necessary, give one or more references to most questions.

Multiple Choice Questions

The use of multiple choice questions is common in examinations. They are composed of one question with multiple possible answers, including one or more 'correct' answers and several incorrect or 'less-correct' answers. We have sometimes noted the number of answers we view as acceptable, and occasionally use an 'all of the above' answer. In general, we have given the logic of the answers with our 'correct' choice. Hydrology is a big and complex field and in writing these questions we were aware of presenting a 'grey' world in black and white. We would encourage students to discuss such issues with your lecturer. We have grouped the questions in Chapters 1, 3, and 4 to some extent but, in doing so, became aware that often questions could be placed in two or more categories.

Matching Questions

There is generally a theme for each question, relating to some aspect of hydrology and watershed management. Some questions call for putting the correct wording in specific places. Others call for cross-matching or pairing.

Fill-in-the-blank Questions

The fill-in-the-blank questions consist of a missing word (or pair of words) in a sentence. The reader is asked to provide the missing words. Fill-in-the-blank questions demand a high level of thought about the discipline area and what is missing from the sentence.

True or False Questions

In these, readers chose true or false values in response to a statement. As noted above we are presenting a 'grey' world in binary black and white, but we have attempted to justify our answers. We do occasionally move into matters of definition, and we would hope that if a student disagrees, that they tease out these areas themself or in conjunction with their professors.

Image-based Questions

In some cases, images were created to illustrate the answer. In other cases, we have looked through the literature to find a relevant maps or illustrations. We found this both interesting and difficult to produce because we are aware of nothing else like it in the literature. It's sort of like a random walk through hydrology.

Essays and Projects

We complete the book with suggestions for essays and assignments. These can be enjoyable to do and can add greatly to your depth of understanding.

1 Multiple Choice Questions

For most questions, tick answers which you think are 'True' or 'Correct'. This may range from zero to all answers. Some cases will specify the number of answers, and some will ask you to pick out an incorrect statement. It is requested that you think carefully before looking up the answers. Please also be aware that in hydrology differences tend to be 'shades of grey' rather than black and white. We hope that you find such questions a useful vehicle for fruitful discussion with your lecturers rather than outrage that the authors are so incorrect. The answers provide a brief commentary on the logic.

Units

1. **A megalitre denotes a million litres. Which abbreviation is correct for this amount of water?**

   A. ml

   B. MLi

   C. ML

   D. mL

2. **20 ML ha^{-1}, as rainfall, is the same as which of the below?**

   A. 1 mm of rain

   B. 100 mm of rain

   C. 1000 mm of rain

   D. None of the above

© Leon Bren and Patrick Lane 2021. *Key Questions in Hydrology and Watershed Management* (L. Bren and P. Lane)
DOI: 10.1079/9781789249682.0001

3. In the older unit of 'points' of rain, there are 100 points to the inch. One inch is 25.4 mm. Hence 20 points of rain is about equivalent to which of the answers below?

    A. 1 mm

    B. 5 mm

    C. 100 mm

    D. None of the above

Dimensions

4. A fundamental division in science is mass [M], length [L], and time [T]. Most (but not all) physical quantities can be ascribed dimensions such as force [MLT^{-2}], representing the product of mass [M] by acceleration [LT^{-2}]. Which of the following attribution of dimensions to hydrologic variables is correct?

    A. Rainfall intensity [L^3T^{-1}]

    B. Streamflow [L^3T^{-1}]

    C. Rate of evaporation [L^2T^{-1}]

    D. Volume of channel storage per unit length of channel [L^3]

Strahler Ordering

5. One of the statements about Strahler order notation used for classifying waterways is incorrect. Which one?

    A. The order assigned will depend on the diligence of map-makers in showing smaller streams

    B. Major rivers such as the Mississippi, Danube, or the Amazon will be sixth-order or higher

    C. Headwater streams originating on farms are going to be first or second-order

    D. All water courses can be assigned a unique Strahler order

6.   **Which two of the following statements are correct about Strahler ordering of watersheds?**

    A. When a lower-order stream joins a higher-order stream, the order of the resulting stream increases by one

    B. When a lower-order stream joins a higher-order stream, the order of the resulting stream stays the same

    C. When two streams of the same order join, the order of the resulting stream increases by one

    D. By definition, streams of the same order cannot join

7.   **Using Strahler ordering, a third-order stream is joined by two second-order streams and three first-order streams. The new stream is then?**

    A. Second-order

    B. Fourth-order

    C. Third-order

    D. Fifth-order

    E. None of the above

Hydrographic Measurement

8.   **The functions of a weir pond behind a measurement section are which of the following?**

    A. To allow water to approach the measurement section with a minimum velocity

    B. To conserve the biota of a stream

    C. To reduce turbulence at the point of measurement

    D. To reduce the noise of the water outflow

    E. To give the water enough height to allow it to fall free of the measurement cross-section

9. Measurement weirs are a favourite method of gauging small, upland streams. A 'stream-rating' is a 1 to 1 relation between flow through a measurement weir and height of water above the notch. Many factors may lead to a deviation from such a relationship. One in the list below should not affect the rating – which one?

    A. Water temperature

    B. Sediment in the water

    C. Aeration status of the water

    D. Changes in the weir pond

    E. Changes in the stream channel below the weir

10. The concept of a 'hydrologic year' is sometimes used in hydrologic data analysis with the start of the year some-where near the end of summer. Which listed advantages below are correct?

    A. With this division of the year, there is less correlation between what happens in a given year with what happened in the preceding year because the watershed soil moisture is usually close to minimum

    B. It sits better with the annual political cycle of many countries

    C. It allows a better pairing of annual rainfall with annual streamflow

    D. It best matches the annual irrigation cycle in many countries

    E. Judicious choice allows 'meshing' with the financial year of many countries

11. A diurnal variation is a characteristic stream flow variation with a 24-hour cycle. Which of the following statements are correct?

    A. In a non-snow watershed, transpiration by trees will often give a diurnal variation with a day-time minimum and a night-time maximum

    B. Diurnal variations are a major periodic component of the hydro-logic cycle.

C. In a watershed with snow cover, the melting of snow by the sun will give a diurnal variation with a day-time maximum and a night-time minimum

D. Diurnal variations in streamflows are often obscured by the bi-diurnal tidal cycle associated with the passage of the moon over the earth's surface

E. Diurnal variations are only found in first and second-order streams

F. They are sometimes referred to as Diehl variations

12. **Consider water travelling across the earth's surface, with the lower part of the water body in contact with the earth and in equilibrium (i.e. not falling). What is about the fastest documented speed that the water will travel at?**

A. 10 kph

B. 30 kph

C. 70 kph

D. 100 kph

E. 200 kph

13. **Which three of the following are necessary to develop a stream rating for hydrographic purposes?**

A. A method of accurately measuring the volumetric rate of flow of a stream at a given time

B. A good working knowledge of Manning's Equation, its errors, and how it is applied

C. A local measure of water height based on a permanent and accessible reference point

D. A good working knowledge of Bernoulli's Equation

E. A stable cross-section which does not have discontinuities in it

F. A method of smoothing the flow to avoid turbulence

Soil Moisture and Vadose Zone

14.   The soil wetness can be defined as

A. the ratio of the volume of water to the volume of pores in the sample

B. the ratio of the mass of water in the sample to the mass of the dry soil

C. the ratio of the volume of water in the sample to the volume of the sample

D. the ratio of the volume of air to the total volume of a sample

E. the ratio of the (volume of air + volume of water) to the volume of the sample

15.   The energy (potential) of water in the soil is most commonly expressed as

A. energy per unit mass $[L^2T^{-2}]$

B. energy per unit weight $[L]$

C. energy per unit volume $[ML^{-1}T^{-1}]$

16.   The Moisture Characteristic Curve is

A. the relationship between soil moisture and soil texture

B. the relationship between air temperature and humidity

C. the relationship between soil matric potential and soil moisture content

D. the relationship between hydraulic conductivity and soil moisture content

17.   There is more plant available water in a clay loam than in a sand because

A. in a sand it drains away too quickly, while in the clay loam there is a large difference between wilting point and field capacity

B. the sand grains pack together tightly, so the pore space between them is minimal

C. plants are better adapted to grow in clays and make the water available using root exudates

18. **All of the following statements about the measurement of soil moisture levels are correct except for one. Select the incorrect one.**

   A. There are a variety of methods for soil moisture measurement in the top 300 mm of soil and all can be made to work satisfactorily or well

   B. Whichever method is chosen, it is a straightforward task to gain a calibration by destructive sampling and drying to constant weight and comparing this result with the instrument reading

   C. Time domain reflectometry (TDR) and neutron probe methods have their advantages and disadvantages and both can be made to work well for deeper measurements

   D. The possibilities of automatic data logging for deep holes using TDR is extremely attractive

   E. Neutron probes are slow, heavy, and cumbersome and, although they can give good results, using them is something of a labour of love

19. **This device requires an aluminium tube be inserted into the soil by drilling ahead of it and then hammering the tube in. The device is then placed on the tube and a part of it is lowered into the tube. The operator then takes a reading averaged over about a minute. Which of the following devices is it?**

   A. A radiosonde device to measure the porosity of the soil by sound transmission

   B. A neutron probe to measure soil moisture

   C. A TDR device to lower wave guides into the soil for a soil moisture reading

   D. A groundwater sampling device. Groundwater is accessed through small holes in the sides of the tubes

   E. An ultrasonic density measurer. The waves are transmitted into the soil and the velocity of reflectance is proportional to the density of the soil material

20. **This device generates a short electromagnetic pulse in a wave guide and measures the elapsed travel time and pulse refection amplitude. This allows the dielectric constant to be measured, which is correlated with a change in the required quantity. Which two statements below are correct?**

    A. The dielectric constant is proportional to the salinity of the water in which the wave guide is embedded

    B. The dielectric constant is proportional to the moisture content of the soil in which the wave guide is embedded

    C. The dielectric constant is proportional to the pressure on the soil at the wave guide location

    D. The method allows fast and automated measurement with results held in a data logger

    E. The dielectric constant is proportional to various chemical contaminants in the soil

21. **A neutron probe (as commercially purchased) has the following attributes**

    A. It gives a continuous monitoring of soil moisture as a function of time at multiple depths

    B. Its low power usage means that it can be set up and left un-attended for long periods of time

    C. It gives a measure of the amount of hydrogen (and other small atoms) in a medium surrounding an access tube

    D. Its specialised measuring of moisture in a very small element of soil requires careful placement

    E. It allows mapping of the uniformity of the soil along the length of the access tube

22. **The movement and storage of water in soils is strongly dependent on the size distribution of the pores in the soil. Which of the following statements is NOT true?**

    A. Micropores are the finest pores, are mostly within individual soil aggregates, and hold the water that becomes available to plants

    B. Mesopores hold the water that is available to plants

C. Macropores drain quickly

D. Micropores hold water at suctions (negative pressures) greater than wilting point

23. **In a wetting and drying soil, the concept of 'hysteresis' refers to**

A. the water potential–soil wetness relation depends on the recent mechanical history of the soil

B. the water potential–soil wetness relation depends on the aeration status of the water and particularly whether bubbles form at low pressure

C. the water potential–soil wetness relation differs with the recent wetness history of the soil, and particularly whether the soil is wetting or drying

D. the water potential–soil wetness relation is particularly dependent on the osmotic properties of material in the soil

24. **If a soil is saturated and allowed to drain freely for several days it will reach**

A. wilting point

B. pore-filled water content

C. field capacity

D. equilibrium with the atmosphere

25. **The vadose zone is usually taken as the soil area around streams. One of the statements below is incorrect; which one?**

A. The vadose zone is the part of the soil between the phreatic (groundwater) surface and the land surface

B. Water is held by the soil at a negative pressure relative to free water at the same height

C. Vegetation always has trouble extracting water from the vadose zone because of the force with which the water is held by the soil matter

D. Water in the soil is slightly below the pressure of water in the nearby stream

E. Although it is sometimes referred to as the 'unsaturated zone', it is in fact a sub-set of the unsaturated zone

26. **The infiltration rate is initially high into a dry soil**

    A. because the saturated hydraulic conductivity is high

    B. because the water table is low

    C. because the matric potential gradient is high

    D. because the soil is not water repellent

27. **A soil is described as 'having a saprolite layer at 2 m depth'. Which of the statements below is a correct description of 'saprolite'?**

    A. A layer of calcium carbonate material formed from limestone in the vicinity

    B. A layer of highly-reduced organic compounds associated with the phreatic surface

    C. A layer of oxidised and weathered rock with some elements of both soil and rock structure

    D. A tough, impermeable layer of clay beneath the A and B horizons

    E. An ironstone layer associated with alternate oxidation and reduction processes in the soil

Groundwater

28. **The basis of hydraulics and groundwater flow are the 'Navier–Stokes' equations of flow. Derived by mathematicians Navier and Stokes, these are partial-differential equations to describe the flow of incompressible, viscous fluids. Which two of the statements below would be viewed as correct by most practitioners?**

    A. When combined with initial and boundary conditions, these break down into a system of equations which are easy to solve and provide real insight

    B. These equations are difficult to solve except for special cases

    C. Although acknowledged as underpinning the mathematics of much of watershed hydrology, these equations are not of much direct use for most situations because of the complexity of application

    D. These equations have long been bypassed by more modern equations

29. **A piezometer measures which one of the following?**

    A. Groundwater temperature

    B. Groundwater pressure

    C. Groundwater velocity

    D. Groundwater salinity

    E. Groundwater movement direction

30. **If you are concerned with land-use hydrology then the most common groundwater aquifer you are likely to encounter will be as follows.**

    A. A confined aquifer carrying water under pressure

    B. A confined aquifer that is not carrying water under pressure (i.e. some of the aquifer has air at about atmospheric pressure)

    C. A water-table aquifer in which the upper bounding surface is the water table

    D. Groundwater is irrelevant – it is surface water

    E. The soil water is held under suction and groundwater does not exist

31. **The water table can be defined by which of the following?**

    A. Pressure of the water in the soil being at atmospheric pressure

    B. The height of water in a hole in the soil sometime after digging

    C. The pressure being measured by a piezometer inserted into the soil

    D. The pressure being measured by a tensiometer in the soil

32. **One of the statements below about confined aquifers is correct. Which one?**

    A. Have an overlying and underlying impervious layer

    B. Always carry hot water

    C. Are rare

    D. Are confined to arid zones

    E. Are associated with land settlement ('deflation') from pumping

33. **An aquiclude is a feature of confined aquifers. Which one of the statements below is correct?**

    A. Is the same as an aquitard

    B. Forms the upper layer of a confined aquifer

    C. Forms the lower layer of a phreatic aquifer

    D. Is the conducting layer of an aquifer

    E. Is the sealed lateral end of a confined aquifer

34. **The product of the aquifer thickness times the hydraulic conductivity is sometimes called the aquifer transmissivity. One statement is 'True' while the others are generalisations which sometimes apply. Which one is true?**

    A. Is constant on a watershed slope

    B. Has the dimensions of $[L^2T^{-1}]$

    C. Is a simple and useful parameter for applied watershed studies

    D. Is only valid for confined aquifers that are fully saturated

35. **A phreatic or water-table aquifer is the common aquifer type sustaining headwater streams. Two statements below are incorrect; which ones?**

    A. The aquifer is bounded by an aquitard at the upper layer and an impervious bedrock at the lower layer

    B. The most common point of intersection of the aquifer and the land surface is along streams

    C. The upper 'transmitting' layer is bounded by the water table

    D. Aquifer material below the level of the stream plays no role in transmission of water to the stream

36. **Karst watersheds, in scientific hydrologic terms are famous for which of the following?**

    A. Their spectacular scenery

    B. The hardness of the limestone rocks found in these spectacular landscapes

C. The extraordinary physical purity of their water because of the filtering action of the limestone rocks

D. Their underground hydrology which, often, has little relation to the ridges and valleys on the surface of the watersheds

37. **Caves and chambers in karst landscapes have been found with submerged depths in excess of 600 m. For comparison, this is well in excess of the 'crush-depth' of most submarines. From the list below, suggest possible consequences of having groundwater at such high pressures.**

A. Just the way it is, no practical consequences

B. Groundwater springs bubble to the surface with unusual force on hill slopes

C. Confined aquifers large distances away, in the absence of leakage, may be maintained at high pressure

D. Groundwater-dwelling bacteria in this circumstance are highly specialised for their cells to withstand such high pressures

E. Sudden reductions in hydrostatic pressures by draining may lead to destructive release of internal pressures within the aquifers (landslips, slumping, etc.)

38. **Deep groundwater is usually in a highly 'reduced' (i.e. low oxygen levels) state. Which one of the following statements is most usually true of shallow groundwater?**

A. In an oxidised state because of proximity to the surface

B. In an oxidised state reflecting its rainfall origins

C. Impossible to be specific because of its variable origins

D. In a reduced state because of lack of oxygen at large soil depths

E. In an oxidised state because of the action of tree roots in adding oxygen

39. **A quanat is a slightly sloping tunnel drilled from a valley into a hillside until it intersects water-bearing rocks. Which two of the following statements are <u>false</u>?**

A. They are commonly encountered in Iran and water elaborate gardens

B. Of great antiquity, they were built by slave labour and cannot be rebuilt

C. They form a self-flowing source of groundwater to sustain farms and gardens

D. At one stage they were viewed as an epitome of sustainability until the cost of building these in human lives was quantified

E. They are an applicable solution to many of the water issues of the Western world

Water Quality

40. **Undisturbed forested natural upland landscapes usually produce high quality runoff water because**

    A. the vegetation protects the soil from erosive runoff and rainfall, which would otherwise mobilise sediment and pollutants

    B. the soils are often highly permeable because of the biological activity and high soil carbon levels which create good structure and infiltration properties

    C. water that does not become evapotranspiration travels to the stream via subsurface pathways through the soil, which slows the velocity of the flow and acts like a big filter

    D. all the above answers are correct

41. **Unsealed roads are a potential source of pollutants delivered to streams due to (choose the best answer)**

    A. high runoff generation, continuous disturbance by traffic, high connectivity to streams, permanent feature of the landscape

    B. high runoff, efficient drainage into streams, the presence of toxic materials from vehicles such heavy metals, oils, etc., permanent feature of the landscape

    C. the presence of contaminants in the materials imported to construct the road surface

    D. none of the other answers are true

42. **Which of the following could not be estimated from a linear relationship between total suspended sediment (TSS) and turbidity?**

    A. TSS load

    B. Phosphorus concentration

    C. Heavy metals attached to clay particles

43. **The reasons wildfire can have such a significant impact on water quality and water supplies are**

    A. fires removes the protective vegetation and causes phosphorus to become more available on exchange sites

    B. fires exposes very large areas of land, the smoke taints the water with hydro-fluorocarbons (HFCs), and fires burn right down to the stream edge, resulting in high connectivity

    C. fires can increase water repellence and runoff, destroy root strength and make otherwise cohesive soil available for transport

    D. fires heats the soil, baking the surface and preventing rainfall from infiltrating, increasing runoff

    E. the heat of the fires can do unusual things to mineralised soil, allowing usually immobile minerals to dissolve

44. **Suppose a 'slug' of water-miscible pollutant is tipped into an upland stream. The contaminant will**

    A. more or less travel with the current, being diluted by turbulence and diffusing into clear water

    B. immediately disperse into the stream and travel downstream as effectively a band of pollutant

    C. remain at the point of entry to the stream

    D. travel along the bed of the stream as a large 'bubble' of pollutant, slowly moving downstream

45. **Conductivity of water is easily measured. This is commonly used as a surrogate for which water-related variables?**

    A. A measure of the ability of water to ionise under electrical current

    B. A measure of the dissolved oxygen content of water

C. The ability of a stream channel to carry boat traffic

D. A measure of the salt content of water

E. A measure of the temperature of water

46.   A 'conservative' (or 'extensive') measure of water quality is one in which conservation of mass applies. Thus, if the volume of streamflow is known, an estimate of the quantity involved can be made. Which of the following measures of water quality are 'conservative'?

A. Stream temperature (°C)

B. Stream turbidity (NTU)

C. Sediment concentration (mg L^{-1})

D. Presence/absence of various types of algae

E. Biomass of algae (mg L^{-1})

47.   Which of the following water quality measures can be multiplied by a volume of water to give a mass of material?

A. Water colour (Hazen units)

B. Turbidity (NTU)

C. Iron (Fe^{++}) concentration (mg L^{-1})

D. Iron (Fe^{+++}) concentration (mg L^{-1})

E. pH level of water

48.   Which of the following water quality measures can be used as an indicator of faecal pollution in a natural waterway?

A. Colour of water (Hazen units)

B. Turbidity of water (NTU)

C. Biological oxygen demand (BOD) (mg of oxygen consumed L^{-1})

D. Total cation content (mg L^{-1})

E. Total anion content (mg L^{-1})

49. Turbidity is often used as a 'catch-all' measurement of water quality in natural streams. Which two of the following statements about this are correct?

A. NTU stands for normalised turbidity units and refers to the self-calibration of the measuring device in use

B. NTU stands for nephelometric turbidity units and the measurement device uses light scattering from particles (nephelometry)

C. NTU is measured with white light but sometimes infrared light is used and this is called FNU

D. In natural streams you don't need a nephelometer at all; lowering a Secchi disk until the pattern disappears is a practical method

E. Turbidity is an excellent parameter because there is usually a good correlation with other important parameters such as sediment load

50. pH of water is a measure of the acidity or alkalinity of natural water. It is not always a satisfactory variable to measure in pure, natural waters. Which one of the statements below is correct?

A. The measurement is difficult and expensive

B. The buffering of natural water is so low that it is hard to get reproduceable results with different brands of instruments

C. The readings are influenced by air bubbles in the water

D. It does not give much information about the water

51. When relatively low populations of humans populate a watershed, which of the following statements about microbiological levels in the water is commonly correct?

A. There is a dramatic change in the species of bacteria and other organisms in the water

B. There is an immediate increase in bacterial populations of streams

C. Bacterial populations of streams remain relatively constant

D. Bacterial populations decrease because of human hygiene

52. **Human faecal contamination of waterways is a huge health issue for many countries of the world. One of the statements below is quite wrong, while others may be arguable; which one is wrong?**

    A. An investment in sewering cities and towns around the world would solve this

    B. Methods using sewerage systems to dispose of ordure (aka faecal matter) use large volumes of water

    C. There is probably not enough water in many countries of the world to use Western-world sewerage approaches

    D. Chlorination of contaminated water supplies would resolve this issue

    E. Young women are particularly affected by social issues of defaecation where toilet facilities are limited and this sometimes restricts their education

53. **'Black water' sometimes occurs when stream water flows through densely vegetated regions. The blackness indicates which of the following?**

    A. Masses of organic material on the stream bed which absorbs incident light

    B. Large masses of sediment carried by the stream water

    C. Dense filtering of light by overhead vegetation, making the stream look black

    D. Tannin compounds leached from decaying vegetation

    E. Large masses of particulate organic matter carried by the stream water

54. **What is the most likely cause of the eutrophication of a lake?**

    A. Low pH

    B. High sediment concentration

    C. High nutrient concentration

    D. Low dissolved oxygen

55. **The 'bed-load' of a stream refers to one of the list below. Which one?**

    A. The depth of particulate matter on a stream bed

    B. Particulate matter that more-or-less rolls downstream along the bed of a stream

    C. The weight of sediment per unit area (pressure) on a stream bed

    D. The weight of sediment matter passing a stream cross-section

    E. The downward force exerted on a lineal metre of stream bed by the weight of water

56. **True story. A student came to see us worried that the long sequence of stream turbidity data sent to him by a government agency was 'funny' – amongst other things it had a clear discontinuity in the middle. It turned out that two different recording instruments had been used. The first installer put a fixed probe into the stream which recorded a value of 1 NTU. This was checked with a handheld instrument which gave a value of 10 NTU. The installer of the fixed probe accounted for the discrepancy by manipulating the data from the fixed probe by adding 9 to every value (i.e. 1 + 9 = 10). A second installer then installed a replacement fixed probe and found the same discrepancy between the fixed-probe and the hand-held probe values. This installer, however, decided to correct for the error by multiplying the fixed-probe value by 10 (i.e. 1x10 = 10). Which of the following statements is correct?**

    A. Adding 9 to each value makes the recording instrument agree with the hand-held instrument and is correct

    B. Multiplying each reading by 10 makes the recording instrument and the handheld instrument agree and is correct

    C. Neither is correct. If you tick this answer, then what might the correct procedure be?

57. **The beginnings of watershed management. In a survey in 2003 by the magazine *Hospital Doctor*, Dr John Snow (1813–1858) was voted 'the greatest doctor ever'. This particularly reflects**

    A. he was 'up with the times' using innovative techniques of anaesthesia (particularly chloroform) and intensive care medicine

B. he observed (after much case-study) a link between water supply and the transmission of the dreaded disease cholera in London

C. he developed complex theories of epidemiology

D. he espoused a healthy lifestyle and good eating with what would now be viewed as a 'vegan' diet

About Water

58. **One of the statements below concerning the fundamental properties of water is incorrect. Which one?**

A. Water flowing in a stream exerts a strong shear stress on the stream bed, attempting to move it downstream

B. Water weighs about 1 tonne per cubic metre

C. Water wets many substances and this begins a process of chemical attack

D. Water sustains combustion and is rapidly oxidised under the right circumstances

E. Water containing sugar will attract pure water into itself

59. **Water, the basic fluid of watershed management, is best described as**

A. A 'non-Newtonian' fluid, in which shear stress is not proportional to the rate of shear strain

B. A 'Newtonian' fluid in which shear stress is proportional to the rate of shear strain

C. A 'plastic' fluid (aka Bingham fluid) in which shear stress is proportional to the rate of shear strain after a threshold value allows initiation of movement

D. An 'ideal' fluid of constant density and zero viscosity

E. A 'compressible' fluid of variable density

60. **Water is a substance whose solid state floats on the liquid state. Which two of the following statements about this are correct?**

A. No big deal – many other substances share this property

B. If this was not the case, the water at the poles of the earth would freeze solid and polar life, as we know it, could not exist

C. Molecules are more loosely packed in the liquid state than the solid state

D. It appears to be due to 'hydrogen bonding' in the water molecule

61. **Water moves to achieve which of the following?**

A. To minimise pressure on itself

B. To the point of lowest total energy it can achieve

C. To a point of zero pressure gradient

D. To the lowest height it can achieve

E. To the lowest velocity it can achieve

62. **Select the one incorrect statement about water here.**

A. Water aggressively attacks many solids found in nature

B. Water has an unusually high ability to store heat compared to other fluids

C. The hydrogen bonding found in water is often used to explain its anomalous behaviour relative to other fluids

D. As water forms from ice, its density decreases over the first few degrees

E. Water can exert negative pressure and hence be pulled along

About Precipitation and Rainfall Processes

63. **Temperature and rainfall vary dramatically in characteristic ways at the global scale because**

A. the uneven heating of the spherical earth sets up air currents to redistribute the atmospheric heat, and hotter air can hold more precipitable water

B. the cooling effect of the ice sheets at the poles makes it colder and wetter there

C. topography affects the air temperature and the amount of solar radiation reaching the earth

D. rainfall rates are different over and land compared to over the sea

64. **Rainfall associated with a cold front is caused by**

A. advancing cold air behind a cold front forcing warm air to lift and as this moist air cools the moisture condenses as rain

B. advancing warm air lifts above stationary cold air and as it does the precipitatable water is released as rain

C. cold fronts produce a lot of rainfall because they carry water from the sea over the land

D. the cold air descends as it is more dense than hot air, and as it does so it warms and attracts moisture which falls as rain

65. **Many hydrology studies have looked at snow precipitation as a part of the hydrologic cycle. Two of the following observations are incorrect or misleading; which ones?**

A. Leonardo da Vinci commented that 'floods occur at the time when the sun melts the snow on the high mountains'.

B. The concept of snow ripening refers to a darkening of the snow-pack due to accumulated dust and organic matter as it ages. This is very pronounced at lower altitudes and near urban areas.

C. Snow density varies from 0.06 gm cm^{-3} in calm conditions to 0.34 gm cm^{-3} in gale-force winds. For most conditions a snow density of 0.1 gm cm^{-3} gives accurate results.

D. Many factors influence snow deposition but snow particularly accumulates in small clearings. This is sometimes used to enhance snow yields.

E. Rainfall falling on a fresh snowpack can generate hydrographs distinguished by the speed of the response and the height of the peak flow for the amount of rainfall.

66. **Which of the following best describes a Thiessen polygon?**

A. A Geographic Information System (GIS) term used as a basis of characterisation of closed and open forms

B. A method of weighting rain-gauge contributions in a watershed by imagining each gauge at the centre of a polygon surrounded by other gauges

C. A GIS model used to encompass spatial variation of hydrologic variables

D. A metal form used to hold a hook gauge on a weir to measure water level

E. An early method of working out watershed boundaries before topographic maps were available

67. **How does topography affect energy and rainfall inputs at the earth's surface (choose the 'best' answer if there is one)?**

A. Aspect affects the solar angle, altitude affects temperature, and topography can affect meso-scale rainfall patterns

B. Topography is a major driver of the global-scale air currents that redistribute heat from the equator towards the poles, which in turn affects the rainfall rate

C. Topography affects the local wind, which determines where clouds form and how much shading occurs over the landscape

D. Topography does not affect energy and rainfall inputs; these inputs are controlled at much larger scales by global circulation patterns and latitude of the location

68. **Orographic rainfall occurs when**

A. incoming clouds approach a mountain and result in rainfall on the downwind side of the mountain

B. an approaching warm front rides up over colder air, and as the air cools it forms rainfall

C. approaching moist air is forced upwards by a blocking mountain and the moisture in the air condenses as rain as the air cools

69. **A unique property of cloud forests is**

A. they are located only at high elevations above the clouds

B. they contribute to the formation of clouds via high levels of stomatal conductance

C. they are generally energy limited

D. they derive a substantial proportion of their water inputs via lateral interception of clouds or mist

E. that although they are often found at high elevations, there are exceptions, and they are not always energy limited

70.   **Cloud forests are most likely to exhibit**

   A. high rates of throughfall and streamflow

   B. high rates of ET and streamflow

   C. high rates of interception and transpiration

   D. high rates of fog drip and soil evaporation

71.   **Droughts tend to be an absence of precipitation. Which two of the following statements are untrue?**

   A. It can be difficult to define when a drought has started

   B. Equally, it is often difficult to define when a drought has stopped

   C. Droughts occur at regular and fixed intervals

   D. What would be viewed as a drought in one area is viewed as 'normal' in others

   E. Droughts have been shown to be an unnatural consequence of climate change

72.   **Rainfall is commonly measured in millimetres (mm). Area is measured in hectares (ha). Water volume is often measured in megalitres (ML) (10^6 litres). Which line of arithmetic below is correct?**

   A. Rainfall of 20 mm in an hour adds 20 ML of water to a 15 ha watershed

   B. Rainfall of 20 mm in an hour adds 3 ML of water to a 15 ha watershed

   C. Rainfall of 20 mm in an hour adds 30 ML of water to a 15 ha watershed

73.   **Precipitation can fall as rainfall or as snow, and the proportion of rain to snow is changing in many places. This is important because**

   A. snow is stored for longer at the surface, changing the timing and magnitude of streamflow events

   B. snowpacks can impact on the infiltration rate of the soil, creating 'rain on snow' events resulting in flooding

   C. snow affects the albedo and rainfall does not

   D. all of the above

74. **You are involved in a study of the rainfall on a forested watershed. Which of the following options would be the best location for a 'representative' rain gauge?**

   A. Under the forest canopy and make a 10% allowance for the interception loss

   B. In a clearing in the forest such that there is a 60° from the horizontal cone of clearance

   C. In a clearing in the forest such that there is a 30° from the horizontal cone of clearance

   D. On a tower some metres above the forest canopy

75. **When considering the effects of climate change on rainfall and evapotranspiration, the most correct statement would be**

   A. both rainfall and evapotranspiration are affected by climate change to different degrees in different places

   B. both rainfall and evapotranspiration are increasing similarly around the globe

   C. rainfall is increasing at the equator, and evapotranspiration is increasing everywhere

   D. both rainfall and evapotranspiration are decreasing everywhere

Watershed and Water Balance Equations

76. **What proportion of the sun's incoming energy is absorbed by the combined atmosphere and land surface (land and sea)?**

   A. 75%

   B. 50%

   C. 25%

   D. 15%

77. **The properties of annual rainfall distributions have come under renewed scrutiny as a result of climate change studies. When plotted as a histogram, a century or more of annual rainfall values commonly approximates a normal distribution. Two of the statements below are incorrect (or perhaps arguable); which two.**

    A. Given the good fit of a normal distribution, it follows that each year of rainfall is an independent event and is not influenced by the preceding year's rainfall.

    B. There is commonly some sort of 'internal structure' within such sequences suggesting that wet years are more likely to be followed by wet years and dry years are more likely to be followed by dry years.

    C. A common tool of analysis of Case B is called 'crossing theory' in which the number of times a cumulative plot of deviation of annual rainfall from the mean crosses zero is compared to that from a true normal distribution with the same mean and variance.

    D. Although a normal (Gaussian) distribution may appear to give a reasonable fit, there are better distributions allowing more accurate prediction.

    E. 'The law of averages' suggests that a wet year should be followed by a dry year to 'maintain the average'.

78. **How might a tree that is energy limited most of the year respond to seasonal water limitation?**

    A. Regulate stomata and drop leaves

    B. Develop smaller xylem vessels

    C. Store water in the root zone by reversing hydraulic gradients

    D. All of the above

79. **In very cold climates, trees need adaptations to deal with a range of temperature-related hydrology issues. Which of these options is most true?**

    A. Trees have structural and mechanical properties to minimise the breakage of branches laden with heavy snow

    B. Trees have special adaptations to prevent damage due to water freezing in the xylem

C. Trees have a range of strategies to prevent water from freezing within living cells

D All these are true

80. The Bowen Ratio is the ratio of

A. radiation to vapour pressure deficit

B. vapour pressure deficit to surface temperature

C. sensible heat exchange to latent heat exchange

D. sensible heat exchange to vapour pressure deficit

81. In 1948 Howard Penman published his seminal equation for evaporation from an open water body. Which of these were not variables in this equation?

A. Proportion of cloud cover

B. Wind speed

C. Vapour pressure deficit

D. Psychrometric constant

82. A plant is well supplied with water, but transpiration rates are low. Which is least likely to influence this?

A. A high vapour pressure deficit

B. A low vapour pressure deficit

C. A large storm event

D. It being a small plant

83. The 'combination' approach to estimating evaporation combines which two physical processes?

A. Evaporation from the soil and from the vegetation

B. Energy balance and mass transfer

C. Advection and latent heat exchange

D. Humidity profile and water supply rate

84. **Q.78 referred to a tree that is energy-limited most of the year responding to seasonal water limitation. In which of the following environments might you commonly find such a tree?**

   A. In very high latitude boreal forests

   B. In high mountain forests in cool-temperate zones (e.g. Eastern Australia, Eastern USA)

   C. In hot desert environments

   D. In cold-desert environments

   E. In coastal forests on the windward side

   F. In coastal forests on the lee (downslope) side.

85. **The second law of thermodynamics is relevant to understanding change in soil and water systems because**

   A. it is a fundamental physical law and it is important for all physical systems

   B. it suggests that changes should only take place that lead to a net increase in entropy

   C. it explains how energy in ecosystems can be converted from one form to another, but never destroyed

   D. it describes the 'change in heat', or the thermal dynamics of ecosystems of soil and water as the short-wave radiation from the sun heats them up

86. **The work of Mikhail Budyko (1920–2001) was concerned with**

   A. the mechanism of the passage of water through the xylem of trees during transpiration

   B. the heat balance of the earth's surface and the conversion of heat to evapotranspiration

   C. the influence of the Coriolis force on air movement

   D. the effect of air viscosity on movement of water through a mixed soil-water-air system

   E. the influence of atmospheric turbulence on the formation of thunderstorms

87. **In 1965 John Monteith produced a crucial addition to the Penman Equation to enable evapotranspiration to be estimated. This was**

   A. the concept of potential evaporation

   B. canopy and aerodynamic resistance terms

   C. the Bowen Ratio

   D. the Boltzmann Coefficient

88. **In forming a watershed water balance, which two of the following would usually be directly measured (or perhaps, assessed from available records)?**

   A. Streamflow

   B. Rainfall

   C. Evapotranspiration

   D. Interception

89. **The concept of a 'sealed watershed' is sometimes laid out as a criterion for watershed selection in hydrology research. Which two of the statements below explain this concept?**

   A. The measurement weir is well constructed so that all stream water at that cross-section passes through it

   B. By routine measurements, a perfect water balance can be achieved

   C. That water can only leave the watershed by evapotranspiration or by streamflow

   D. That there is no subterranean 'groundwater' transfer to other basins

90. **By the water balance of a watershed we mean**

   A. sum of inputs of water = sum of outputs of water

   B. the streamflow emanating from the watershed

   C. the evapotranspiration from the watershed

   D. the deep seepage loss from the watershed

91. **Evapotranspiration is**

   A. the water loss from bare soil to the atmosphere

   B. the difference between rainfall and streamflow

C. the water stored in plant leaves

D. none of the above

92. **In which country was the concept that the rainfall has the capacity to sustain the flow of major rivers first demonstrated to the satisfaction of scientists?**

A. USA

B. UK

C. Germany

D. Austria

E. France

93. **Given that P = annual precipitation depth, E = annual evapotranspiration, Q = annual streamflow, and Error = error (unmeasurable variables), the annual water balance for a small watershed is usually expressed as**

A. $P + E = Q - Error$

B. $P - E = Q + E + Error$

C $P = Q - E + Error$

D. $P = Q + E + Error$

94. **A trial watershed project routinely measures precipitation and streamflow. Which of the following statements are correct?**

A. Evaporation is such a minor component of the water cycle, it is not worth measuring directly

B. Given that evaporation is not being measured, we really can't say anything about it

C. The difference between rainfall (expressed in mm) and streamflow (expressed in mm) is a measure of the evaporation, deep seepage (watershed leakage), and measurement errors

D. Deep seepage and measurement error are usually so small that the difference between rainfall and runoff is a good measure of evaporation

95. **Thinning of forests is a classic way of 'enhancing' water yields from forested watersheds. Which one of the following statements about this is not correct?**

A. Thinning reduces the water use of the forest by reducing the amount of sapwood which passes (transpired) water

B. In areas of lower rainfall, thinning does not give a marked response because the retained forest may still use most or all of the water available

C. An aggressive method of thinning is to cut strips through the forest. This works quite well but the appearance of the forest may not be acceptable on other grounds

D. The effects of thinning are evident in enhanced water yield from the forest for many, many years

E. After thinning the vegetation responds so that after a few years the water use of the forest returns to the pre-thinning level

96. **A 100 ha watershed has an annual rainfall of 1500 mm, of which half appears as streamflow. Water in a nearby town is sold to consumers at $3 per KL. Which of the following statements are correct?**

A. At 15 ML ha^{-1} rainfall, on 100 ha, then the watershed produces 750 ML of water annually

B. Given the urban price of $3 KL^{-1}, this water could be valued at $2.25 million by simple arithmetic

C. The valuation of this stream output is $2,250,000

D. Since the watershed has produced this amount of wealth, the watershed owner should be paid a proportion of this amount

97. **In an undisturbed temperate forest landscape, which would we expect?**

A. High evapotranspiration, good streamflow, low peak flows

B. Low evapotranspiration, low streamflow, high peak flows

C. Low evapotranspiration, high streamflow, low peak flows

D. High evapotranspiration, high streamflow, high peak flows

Plant Evapotranspiration and SPAC

98.   'SPAC' (or 'Surface-Plant-Atmosphere-Continuum') is sometimes used to characterise the evapotranspiration concept in watershed hydrology. Which of the following statements about SPAC are correct?

A. SPAC shows an energy gradient in which the water at the top of a tree is at a much higher energy/pressure level than elsewhere because of the tree's height

B. The continuum part means that the water being carried through SPAC always has an energy/pressure gradient (i.e. there are no discontinuities in water pressure until the water actually leaves the plant leaves)

C. The driving force of SPAC is the energy of the sun in keeping water in the atmosphere at a very low energy/pressure level

D. SPAC characterises very well the pumping and pulsing actions of tree roots in getting water to the top of tall trees

99.   The potential evapotranspiration rate is

A. the rate that water evaporates from an open body of water

B. the rate of evapotranspiration from a well-watered field if wind speed, temperature, and relative humidity were at an optimum

C. the maximum rate of evapotranspiration possible for a particular temperature

D. the rate at which evaporation and transpiration would occur from a large area covered in growing vegetation with unlimited access to soil water

100.   It is often hypothesised that leaf area index (LAI) of vegetation is correlated with evapotranspiration capacity. LAI is

A. ratio of winter leaf area to summer leaf area

B. ratio of leaf area to (horizontal) ground area

C. ratio of leaf area to stem cross-sectional area at ground level

D. ratio of upper canopy leaf area to lower canopy leaf area

E. probability that a vertical projection from the ground will pass through one or more leaves

101. **The leaf area index of a forest or crop can be measured by which of the following methods?**

    A. A specialised device called a porometer

    B. Injecting radioactive tracers into tree stems and seeing which leaves show their presence

    C. Harvesting leaves from crowns and measuring their area

    D. By measuring the probability of raindrops passing through the forest canopy

    E. Using a specialised device called a ceptometer which measures photosynthetically active radiation

    F. Analysis of LIDAR (light detection and ranging) data

    G. Analysis of hemispherical and crown cover photographs

102. **Enhanced CO_2 levels give a carbon fertilisation effect. This can lead to increased water use efficiency and faster plant growth, but this is not always the case, because**

    A. sometimes other factors are limiting water use and growth, such as nutrients

    B. some plants are already growing at a maximum rate

    C. water use efficiency will only increase where soil organic matter is not limiting

    D. this effect does not occur in all geologies

103. **Which list has the potential of water in the correct order, going from the least to the most negative?**

    A. Atmosphere, leaves, xylem in branches, xylem in stem, roots, soil moisture

    B. Xylem in stem, soil moisture, leaves, atmosphere, xylem in branches

    C. Soil moisture, roots, xylem in stem, xylem in branches, leaves, atmosphere

    D. Soil moisture, xylem in stem, roots, xylem in branches, atmosphere, leaves

104. **The reason plant water use efficiency has been found to increase with climate change is because**

A. the efficiency of rainfall is increasing with a warming atmosphere that can hold more water

B. the higher temperature means that plants can transpire more efficiently

C. the higher CO_2 concentrations means that less water is lost from the stomata for every unit of carbon that enters

D. higher wind speeds mean that evaporation from stomata is more efficient

Satellite Imagery and Remote Sensing Hydrology

105.   The acronym DEM refers to which of the following?

A. Differenced Energy Model

B. Discrete Evaporation Model

C. Digital Elevation Model

D. Direct Evapotranspiration Measurement

E. Domained Excel Model

106.   'LandSat' was an early satellite monitoring the earth's surface. Which one of the following statements apply?

A. Images may be useful but are tightly held by military requirements

B. The satellite is viewed as a worthy but primitive venture into this area

C. This satellite and updates gives the longest continuous space-based record of the earth's surface and have been used in many hydrologic studies

D. The images were found to be of little use in hydrology

107.   Which one statement about satellite imagery would be accepted by most scientists?

A. They have rendered ground-based hydrology monitoring obsolete

B. They allow direct and accurate measurement of evapotranspiration

C. They have provided complementary (but sometimes difficult) tools to supplement other methods

D. They are not of much use in watershed hydrology because there is little ground penetration

E. They are enigmatic and require such specialised skills that they are of little use to anyone other than the military

108. **NDVI is often computed for each pixel in satellite-based hydrology studies. Which of the following statements below about NDVI are correct?**

A. It stands for 'Nominal density value indicator'

B. It stands for 'Normalised difference vegetation index'

C. It is calculated as the ratio (NIR-Red)/(NIR + Red) where NIR is near-infrared reflectance and Red is red reflectance

D. Because of the dominance of red in arid landscapes, it is particularly effective at indicating iron-dominated landscapes and hence widely used in mining evaluation

E. It is particularly effective at distinguishing healthy, green vegetation from other land covers

General Hydrology

109. **One of the following attributes of big dams is incorrect; which one?**

A. They provide regulation of flow (for better or worse)

B. They tend to make the water immediately below the dam free of sediment

C. They enhance the 'natural values' of rivers and streams

D. The obstruct fish passage on the river

E. They tend to be objects of pride for communities

110. **'Imbrication' of stream beds commonly refers to which of the list below?**

A. The arrangement of stream-bed material in a riffle and pool sequence

B. The stream structure sometimes found due to fish burrowing in stream beds

C. The overlapping structure (e.g. as found on roof tiles or weather-boards) of components such that there is only a low resistance to downstream flow

D. The occurrence of 'holes' in the stream bed at the foot of plunge pools

E. The essentially random placement of components such that the stream bed is homogenous

111.   In a flooding forest, a fundamental parameter is the hydroper-iod. Which one of the definitions below defines this?

A. The months in which the forest is expected to be dry (unflooded)

B. The frequency with which the forest floods, expressed as the probability of flooding in a given year

C. The probability of flooding in a given year

D. The months of the year over which the forest is expected to be flooded

E. The number of months in a year in which the forest might reasonably be expected to be flooded

112.   Which of the following statements about small coral islands of the South Pacific are correct?

A. Their heavy rainfall ensures that fresh water is plentiful

B. By the nature of such islands, the underlying rock is solid and impervious

C. The fresh groundwater is a 'lens' recharged from above, stored in porous aquifers, and effectively floating on underlying saline groundwater derived from the ocean

D. Fresh groundwater is usually both scarce and limited

E. Groundwater is not a concept that is applicable in such a situation

113.   Arid zone vegetation often displays distinct patterns. What is the most likely reason?

A. Groundwater patterns

B. Variability in soil texture

C. Preferential trapping of rainfall or dew by the vegetation

D. Ancient root systems from wetter periods

114. **Older texts on hydrology sometimes appear obsessed with the concept of overland flow – water passing across the land surface to the enter a stream network. Which of the following statements are correct?**

A. 'Overland flow' has been replaced by the concept of 'near-surface flow' in many books and papers

B. Many surfaces are too rough and the infiltration capacity too great to sustain true 'overland flow'

C. Overland flow, when it occurs (usually in urban areas due to paving) can give spectacular flash flooding

D. A benefit of agricultural and forested land in watersheds is that it reduces the occurrence of overland flow

115. **When a watershed is severely burnt, hydrographs may exhibit the following?**

A. Loss of the diurnal variation due to death of the transpiring tissue

B. Enhanced soil water repellency

C. Stimulated plant growth because of the release of nitrogen

D. Drying up of the stream because of boiling-off of the water by the fire

E. Enhanced streamflow sometime after the fire because of the loss of transpiration

F. Extremely high and erosive 'spike flows'

116. **The Unit Hydrograph is the hydrograph generated**

A. divided by the depth of rainfall

B. per unit of watershed area

C. by a unit amount of rainfall in unit time

D. by an 'instantaneous spike' of rainfall of unit quantity

E. by a unit volume of overland flow

117. **'Flood routing' refers to**

    A. Using the square root of the flow velocity to compute flood wave speed

    B. Estimating the probability of downstream erosion from a flood

    C. Computing the time locus of floods along a channel network

    D. Working out the channel network that floods will travel down

    E. Evaluating the probability of downstream structures exceeding their maximum permissible flow

118. **Many mountain streams show a systematic variation of longitudinal cross-sectional structure with deep, still water alternating with 'rapids' or 'overflows'. This structure is**

    A. difficult for biota because of the low oxygen levels in the pools

    B. a consequence of the accumulation of erosion debris and poor land management

    C. a non-random, natural 'pool-riffle' structure with a reasonably consistent wavelength

    D. due to erosion associated with removal of larger trees from the banks

    E. formed by occasional, extremely high flows associated with pioneer activities upstream

119. **Watershed evolution is the theory that long-term energy and water inputs leave an 'evolutionary footprint' on current-day hydrologic functioning. Why might this be useful to 21st century hydrologists? There may be more than one correct answer.**

    A. We can better understand the potential impact of climate change

    B. We can forecast the effect of management practices on the system

    C. We can use this knowledge to increase production

    D. We can better understand what might occur when the system is subject to natural systems

120. **'Saltation' is a hydraulic process described by which one from the list below?**

    A. The tendency of dry salt in arid lands to be blown by the wind

    B. The accumulation of salt scalds at the toe of a slope

    C. The 'jumping' of sediment particles along the bed of a stream

    D. The tendency of heavier saline water to gravitate to the bottom of aquifers

    E. The accumulation of more saline water in stratified pools in streams

121. **Landslides are a well-explored hydrologic hazard. The lists below shows factors often associated with landslides involving movement of soils and underlying rock. One factor stands out as predisposing landslides – which one?**

    A. Heavy rain leading to wet hill slopes

    B. Lack of tree roots binding the hill slope together

    C. 'Slippery' layers in the rock or soil

    D. Recent changes in the distribution of water on slopes (roading, etc.)

    E. Additional weight placed on slopes (e.g. tree growth)

    F. Removal of 'toe support' for the slope

    G. Earthquakes or earth-tremors causing soil liquefaction

122. **Throughout history there have been many 'great' (aka disastrous, killing, tragic) floods. Which one of the following has not, to the our knowledge, been associated with a 'great flood'?**

    A. Torrential rainfall

    B. A mountainside collapsing into a dam and displacing the water in the dam

    C. An extremely rapid thaw of snow on a mountain

    D. An earthquake liquefying soils and leading to release of water from the soil and catastrophic downstream flooding

    E. A volcano 'blowing up' a lake, with the disruption flooding downstream

123. **Which of the descriptions below describes the geometry of a common rainbow formation as you stand looking at it, facing its centre?**

   A. Sun out and to your right

   B. Sun out and at your back

   C. Sun in and at your left

   D. Sun out and at your left

   E. No fixed geometry – to a first approximation, random in space

124. **In ecohydrology, 'optimality' expresses the idea that**

   A. ecosystems are all optimised with respect to their surroundings

   B. ecosystems have a theoretical optimal state, and the ecosystem is constantly adjusting its state on a trajectory to this optimum

   C. ecosystems are never optimal because energy conversions between the trophic levels are inefficient

125. **When rainfall leads to a moderate streamflow storm hydrograph, the water entering the stream from a forested watershed slope is mainly**

   A. the rainwater ('new water') falling on to the watershed

   B. water stored in the slopes (and/or groundwater domain) from a previous storm ('old water')

   C. rainwater from the storm that has passed through the watershed soil

   D. rain water passing down the soil surface into the stream

126. **Which of the following statements about the 'variable source area concept' in watershed science are true?**

   A. The concept was introduced because the dogma of overland flow contributing to forested streams was found to be mostly inapplicable

   B. The concept goes back to the seminal writings of Leonardo da Vinci

   C. The concept is that the parts of the watershed contributing water to the stream varies from storm to storm

D. The concept was devised by US forest hydrologist John Hewlett

E. The concept is that, in a storm, the area of watershed contributing water to the stream expands over the course of the storm and contracts when the storm ends

127. **Two of the following statements about the River Nile are incorrect; which two?**

A. The Nile has been dammed by the Aswan Dam in Egypt, with generally beneficial effects

B. The Egyptians developed elaborate structures called 'Nilometers' to record the water levels

C. Because of modern technology the Nile River is far less important to Egypt than it used to be

D. The ability to predict the volume of coming inundations was part of the mystique of ancient Egyptian priesthoods

E. The origins of the Nile were unknown to European civilisation until the mid-19th century

128. **The 'Hurst Effect' refers to**

A. the tendency for water moving into the soil to 'finger' out

B. a small positive correlation between streamflow in successive years

C. the tendency for sediment to move to the centre of a vortex

D. the fractal nature of long-term streamflow records over time

E. the tendency for histograms of hydrologic data to be positively skewed

Modelling and Quantitative Hydrology

129. **Occam's razor is**

A. a sharp tool used to cut back models

B. the original scooter used by the 14th-century philosopher William of Occam

C. the idea that models should be as simple as possible

D. the concept of making sure everything is included in a model

130. **A Monte Carlo simulation is a way of modelling that involves**

A. running a model over and over using distributions of parameters as inputs

B. combining an empirical and a deterministic model to generate distributions of outputs

C. multiplying together input distributions to determine an output distribution

D. simulating the values of the input parameters using an empirical model

131. **Three of the most important things to consider when constructing a model are**

A. calibration, testing, model range

B. cost, time availability, skills

C. resolution, parameterisation, testing

D. purpose, resolution, resources

132. **A physically based hydrologic 'model' usually consists of one or more equations, an initial state, boundary values (i.e. what is happening at the edge of our domain), and some input of change over time. Classify the statements below as 'True', 'Maybe True', or 'False'.**

A. The watershed ridge is a boundary and there is usually no inflow from this unless water is being imported by a channel

B. The domain is always a rectangle of the land surface

C. For a physically based model, the parameters can, in principle at least, be measured by an appropriate laboratory or field setup

D. The initial values are usually obtained by field measurements

E. The spatial arrangement of components is not usually a factor in the model

133. **'Topmodel', developed in 1978, was considered to be a significant advance in watershed modelling. Was it because it**

A. considered the role of weathered rocks in storing water

B. considered the role of topography in routing water

C. explicitly solved for stream routing during storms

D. it was an early computer model

134. **Consider a paired watershed experiment of four basins measuring daily rainfall and water outflow, to last for 50 years. Basin 1 will be left untouched, Basins 2 and 3 will have the same land-use treatment ('Land-use 1') applied after a 5-year 'calibration period'. Basin 4 will have a second land use ('Land-use 2') imposed after the calibration period. Class the following statements as 'Correct', 'Maybe', or 'Incorrect'.**

A. Such a design is at the highest levels of experimental methodology in science

B. Such a design is at the highest level of experimental methodology in watershed science

C. Basins 2 and 3 are replications of the treatment

D. Basin 1 forms a control for the treatments of Basins 2, 3, and 4

E. The control gives a classic 'no change' reference for the 50 years of experimentation

F. A 'calibration period' developed over the non-treatment 5 years, will greatly enhance the experimental analysis

Hint. To visualise such a project it is suggested that students sketch a diagram of such a project using watersheds each of about 50 ha.)

135. **Hydrologic data is often positively skewed. This means which of the following?**

A. The mean and variance are close to those given by a normal distribution

B. There is a long 'tail' of data to the left of the mean

C. There is a long 'tail' of data to the right of the mean

D. The mean and variance are the same

E. Logarithmic transforms of the data are sometimes used to make the distribution closer to that of a normal distribution

136. **One of the most common empirical hydrologic models is called 'The Rational Model'. It is embedded in many engineering applications and is written as Q = KCIA in which A is watershed area [L^2] and I is the rainfall intensity [LT^{-1}]. C is the fraction of rainfall converted to streamflow and is dimensionless. K is a unit converter to allow non-homogeneous but convenient units. Which two of the following statements are correct?**

A. If we use the model over long time sequences, C will be a function of time

B. If A is given in hectares, and I is in mm per hour. Then if we want to have Q in units of litres per hour, it follows that K would have to be 10

C. C is often taken as a function of I, with a large I implying a low C

D. For a given I, a forested watershed would have a low value of C while a paved watershed would have a high value of C

137. **Suppose you have flow data from two watersheds, 1 and 2. Accumulation of data is making a running sum. Double-mass plots were an older visual method of hydrologic data analysis and are still useful for visualising hydrologic change. A double-mass plot is which of the following?**

A. Plot of flow from stream A as a function of flow from stream B

B. Plot of accumulated flow in stream A and B as a function of time

C. Plot of accumulated flow in stream A against flow in stream B

D. Plot of accumulated flow in stream A against accumulated flow in stream B

E. None of the above

138. **Which watershed would produce the most flow after the same storm event?**

A. A 5000 ha watershed with high drainage density

B. A 5000 ha watershed with high slope

C. A 5000 ha watershed with concrete channels

D. I need more information

Urban Hydrology

139. **The key hydrologic problem created by urbanisation is (choose the best answer)**

    A. an increase in the frequency and magnitude of peak flows, and a decrease in low flows or base flow, particularly in summer

    B. an increase in the total amount of water coming out of the watershed

    C. an increase in baseflow in winter, and decrease in baseflow in summer

    D. an increase in water pollutants from fertilisers used in gardens

140. **What commonly happens to baseflow as a result of urbanisation?**

    A. It decreases because impervious surfaces prevent infiltration

    B. It increases because of anthropogenic sources, over-irrigation and leakages

    C. It increases because of a loss of vegetation and thus evapo-transpiration

    D. Any of these are possible, depending on the watershed context

141. **What is the typical annual runoff coefficient (C) from an impervious area?**

    A. 0.2

    B. 0.5

    C. 0.9

    D. 1.0

142. **Which of the following are objectives of 'Water Sensitive Urban Design' or 'Sustainable Urban Drainage Systems'?**

    A. Restore more natural flow regimes, by reducing peak flows, restoring more natural baseflow, reducing overall flow volume

    B. Maximising the runoff coefficient of urban watershed

    C. Reducing the loads and concentrations of pollutants

    D. Evacuating water as quickly as possible from the landscape

E. Maximising the amenity of the urban landscape using water to support vegetation

F. Reducing demand on potable water supplies

G. Decreasing the time of concentration of urban watersheds

143. **Connectivity between impervious surfaces and the stream network is important because**

A. for water quality, as disconnecting stormwater-producing surfaces allows infiltration, filtering and trapping of pollutants

B. for peak flow reduction, as infiltrating water into the soil slows down its transmission to the stream channel, and therefore reduces the peaks in the hydrograph

C. for flood protection; it is important to transmit as much of the stormwater to the stream as quickly as possible to maximise drainage and prevent flooding

D. all of these could be true, depending on your perspective and what you value and think is most important in urban hydrologic systems

144. **A road-builder building an unsealed road in a forest near a pristine stream solves one problem, but creates another. This situation could best be described as**

A. the road will bring traffic closer to the stream, increasing the potential for oils and other contaminants to reach the stream, contaminating it

B. the road-base is compacted to shed water and protect the bearing capacity of the road, but this generates high runoff rates that can carry sediment and contaminate the stream

C. the road will intercept groundwater and modify the hydrologic pathways in the watershed, affecting peak flows

D. the road-builder will import materials to construct the road-base, and these will ultimately end up in the stream, polluting it and impacting on the aquatic ecosystem

Land-use Hydrology

145. **A forest in a 700 mm rainfall zone is cleared for cropping. Could the crop have the same ET rate?**

    A. Yes, when averaged over a year and if supplied with water to ensure close to potential ET

    B. No, the crop could not intercept or transpire at the same rates as the trees

    C. Only if they were large plants such as watermelon

    D. Yes if the soil water holding capacity was unchanged

146. **The paired watershed experiment consists of changing the land use on one or more small watersheds and observing the change in comparison with a 'control' watershed which remains in the same state. Which of the answers below would be accepted by most hydrologists?**

    A. The development of paired watershed experiments was a major development in land-use hydrology

    B. Although they are expensive to get established, their maintenance is relatively inexpensive

    C. These 'experiments' can last a long time so whether the 'control' is really a control or a watershed on a journey to somewhere else is arguable

    D. The experimental design would be greatly improved by replication

    E. The experiments are crippled by not having a 'double blind' design in which the data analysts have no idea of the treatment or the experimental layout

147. **A paired watershed experiment is conducted with a control watershed. The flow in the treated watershed was about the same as the control before treatment. After the treatment the flow increases by about 2 ML ha^{-1} for a few years and then, in the next 5 years or so declines to about the same flow as the control. Annual rainfall in the watershed is about 800 mm per annum. The control and treatments were**

    A. native forest control and thinning of native forest on treated watershed

    B. grassland control and planting up of grassland watershed on treated watershed

C. grassland control and clearfall/regeneration of native forest on treated watershed

D. native forest control and clear fall logging of native forest on treated watershed

E. mature plantation control and logging and replanting of plantation on treated watershed

148. **A paired watershed experiment consists of two grassland watersheds in about 1000 mm per annum rainfall. One is then stripped of grass and planted with a pine plantation. Good weed control is implemented and the plantation becomes established. Which of the following responses might you expect?**

A. The flows from the two watersheds would be unchanged

B. The flow from the planted watershed would increase and then remain at a higher level permanently

C. The flow from the planted watershed would rapidly decrease until the stream flow completely stopped

D. The flow from the planted watershed would increase for some years then gradually decline over the next 30 years as the pines 'got a grip'. Flow would decrease by about 200 mm per annum

149. **A paired watershed experiment consists of two watersheds carrying mature native forest in about 1000 mm per annum rainfall. One has the mature forest cleared, the debris burnt and is then planted with a pine plantation. Good weed control is given and the plantation becomes established. Which of the following responses might you expect?**

A. The flows from the two watersheds would be unchanged

B. The flow from the planted watershed would increase by about 300 mm per annum for some years and then progressively diminish but would always be higher (over the 30-year plantation life) than that of the native forest control watershed

C. The flow from the planted watershed would rapidly decrease until the stream flow completely stopped

D. The flow from the planted watershed would remain unchanged from the pretreatment case

150. **A paired watershed experiment has heavy forest on both the control and the experimental basin. Both show a clear diurnal variation in flow in summer. The trees and understorey are removed for 50 m from either side of the stream as the treatment. Which of the following might reasonably be expected?**

A. Flow in the stream would diminish and the amplitude of the diurnal variation in flow would decrease

B. Flow in the stream would diminish and the amplitude of the diurnal variation in flow would increase

C. Flow in the stream would increase and the diurnal variation would disappear

D. Flow in the stream would increase and the amplitude of the diurnal variation would increase

151. **Fourier analysis allows a time-variant waveform such as a streamflow record to be broken into sinusoidal components of varying frequencies. Which of the following regular frequencies might you expect to find in your long-term paired watershed data?**

A. A bi-daily tidal cycle associated with passage of the moon

B. A diurnal cycle (day/night) associated with rotation of the earth

C. An annual cycle associated with the movement of the earth around the sun

D. A 21-year cycle associated with sun-spot growth and decay on the sun's surface

E. A 150-year cycle associated with the joint proximity of Venus and Mars to the earth

152. **Suppose you had a 5-minute sampling frequency on your data logger, and you did a Fourier decomposition of the flow chart. This shows a spike in high frequency variations and the higher the sampling frequency becomes, the more intense the high-frequency Fourier spike becomes. Rank the likely causes from the list below.**

    A. Electronic problems causing transients

    B. Inaccuracies/inadequacies of your flow transducer at a given water level

    C. Tectonic vibration causing minute-by-minute changes in slope outflow

    D. Water turbulence in your measurement structure causing your water level transducer to fluctuate

    E. A watershed with very rapidly changing weather

153. **John Hewlett's (1922–2004) projects looked at the responses of streamflow from forested watersheds to periods of rainfall. These did show some interesting aspects of forest hydrology; which of the below are true?**

    A. Watersheds with thin soils would respond in a similar way to those with deep soils initially, but as soil capacity filled would show a 'violent response' to more water

    B. Watersheds with strong native burrowing-animal activity were far less responsive than those without

    C. There was little relationship between the peak flow achieved by a given rainfall and the maximum rainfall intensity in the rainfall

    D. The streamflow at the start of the rainfall had a positive correlation with the volume of stormflow and the peak streamflow achieved

    E. Watersheds with very large trees had such a major interception loss that little water reached the ground, thereby making them unresponsive

154. **The most common pattern of response to clear falling and regeneration of forest on the slopes of a first-order watershed in temperate forest areas without major snow falls is**

    A. a dramatic increase in stream outflow which persists for many years. This can lead to sustained downstream erosion

    B. no change in watershed outflow

    C. an increase in watershed outflow for about 3 years, followed by a reversion to the historic outflow pattern

    D. a dramatic decrease in streamflow that persists for many years

155. **The long-held theory of the response of streams to insect attacks on forest in their watersheds is that streamflow increases, reflecting the reduction in ET. However, more recent work has indicated which of the following?**

    A. Streamflow decreases because of death of tree crowns led to increased sub-canopy evapotranspiration so that net streamflow decreased

    B. Streamflow decreases because undergrowth stimulated by the increase in exposure transpired more than the trees

    C. No change in streamflow

    D. Streamflow increase

156. **The 'active layer' of permafrost is**

    A. the soil rooting zone in cold climates

    B. a layer of soil, rock or sediment that freezes and thaws annually above the permanently frozen zone

    C. the zone from the lowest depth of annual thawing to the depth at which the geothermal temperature is above freezing

    D. the upper boundary of the permafrost that is actively growing as new water infiltrating from above freezes and causes the permafrost to grow

157. **The 'Dust-Bowl' events of the 1930s in the US Southern Plains were caused by**

   A. desert sediments blown in by high winds

   B. a combination of drought and the destruction of soil structure by tillage for cropping

   C. widespread irrigation failure, leaving dry fields subject to erosion

   D. crop disease that wiped out crops and led to erosion

158. **Cumulative hydrology effects refer to**

   A. the use of accumulated or integral variables (e.g. double mass plots) in assessing hydrologic impacts

   B. the net hydrologic effect of many land-use activities on a watershed

   C. the accumulation of net financial benefits of land-use activities on a watershed

   D. the accumulation of nutrients in a stream as a result of sub-surface drainage

159. **'Cumulative hydrology effects' are sometimes known as 'the aliens of hydrology' because**

   A. they are so potentially damaging to the world

   B. everyone knows they are there, but they are impossible to show

   C. they don't exist

   D. they exist but in a form we can't quantify

160. **The first 'paired watershed' experiment in the world was**

   A. the start of the Coweeta Paired Watershed Experiments in North Carolina, USA (forestry impacts)

   B. Wagon Wheel Gap experiment in Colorado, USA (deforestation impacts)

   C. Emmental Project in Switzerland (land-use hydrology)

   D. Croppers Creek experiment in Australia (plantation impacts)

   E. Mokobulaan experiment in South Africa (plantation impacts)

161. **John Hewlett (1922–2004) was an American forest hydrologist noted for which two of the list below?**

A. An adherence to the theory that overland flow was the dominant source of streamflow in forested watersheds

B. Development of the 'variable source area model' which argues that subsurface processes in a core area of watersheds near streams provide the storm rainfall response

C. Development of a finite-element approach to solving the fundamental equations of soil moisture movement in forested environments

D. Demonstrating the importance of unsaturated 'interflow' in replenishing valley aquifers and maintaining streamflow

E. Quantifying the importance of snow in maintaining streamflow over summer periods

Answers

1. C. Journalists often confuse ml (10^{-3} L) with ML, with ludicrous results.

2. D. 1 ML ha^{-1} is the same as 100 mm of rain. Hence 20 ML ha^{-1} would correspond to 2000 mm of rainfall.

3. B. Close to 5 mm. Even in metricated countries, farmers sometimes still talk in points of rain.

4. B is the only one correct (e.g. flow is given in m^3 per second). Rainfall intensity is LT^{-1} (e.g. mm/hour). Rate of evaporation is LT^{-1} (e.g. mm/day). Volume of channel storage per unit length of channel is L^2 [L^3/L]. Effectively it is the area of cross-section.

5. D is incorrect. A. B, and C are correct. D is incorrect – the ordering system breaks down when streams disappear into the landscape, pass through large lakes, or become braided over a flat landscape. It is, at best, a useful simplification.

6. B and C are correct. A and D are false.

7. C. Third-order. When streams of lower-order join a stream, the order remains unchanged.

8. A, C, and E are correct. A weir will develop its own small ecosystem but it is different to that of the stream, so B is false. A weir is usually a noise emitter so D is false.

9. E. Changes in the stream channel below a weir should not impact on weir rating (if it does, the nappe is said to be 'submerged'). A through D would all impact on the rating by changing the viscosity of the water or impacting on the flow approach to the measurement section to some extent.

10. A and C (C is A restated) are correct. The idea is that major hydrologic events should not be split between 2 years. This still happens but judicious choice of months can minimise this.

11. A, C, and F are true. B is false – they are a relatively small component and often obscured by larger variations such as rainfall, or hidden by insensitivity of measurement systems. D is false – although the moon does give a bi-diurnal variation in the oceans ('tides') there is no evidence that this is visible in stream systems. E is false – they occur in large and small hydrologic systems.

12. C. The scientific literature is unspecific on this question, but when Malpasset Dam (France) suddenly had an abutment failure and collapsed in 1959, the 40 m high wall of water released was accurately clocked as travelling at about 70 kph for some kilometres. It was a huge disaster. Imagine trying to pedal your bike to try to keep ahead of it! Fortunately for us, most waves travel considerably more slowly.

13. A, C, and E would provide a minimum for developing a flow rating. A number of rating points would be measured at different flows and a working relationship developed giving the flow through the cross-section as a function of stage level. This may take some months or years to develop. B and D may be desirable but are not necessary. F may be desirable to reduce error in any measurement but, for large rivers in particular, may not be possible.

14. B and C. B is called the mass wetness, and C is called the volumetric wetness. A is the degree of saturation. D is the air-filled porosity. E is the void ratio.

15. B. This is commonly referred to as 'head'. This makes arithmetic simple and helps visualisation.

16. C. However, A and D are related to the curve parameters via relationships between soil texture, volumetric water content and matric potential. B is unrelated.

17. A. Water may bond to the clay particles better than the sand particles.

18. B is incorrect, particularly for deeper measurement. Problems usually lie in difficulties of sampling and dealing with inhomogeneous materials and macropores.

19. B. A neutron probe for soil moisture measurement. The source is lowered to the appropriate depth and a reading taken. By moving up and down the tube, the whole depth is 'logged'. Sometimes a capacitance device is used similarly but the tube is not aluminium.

20. B and D. The method is Time Domain Reflectometry (TDR). It provides an alternative to the neutron probe for deep moisture measurement if (and it's a big if) one can work out how to insert the wave guides. The method does allow more-or-less continuous scanning.

21. C. It gives a measure of the density of hydrogen in a diffuse sphere of about 150 mm around the access tube. Since water is commonly the main source of hydrogen, this is, *de facto*, a measure of the moisture content. Neutron probes cannot usually be used for continuous measurement and require an operator. Depths have to be logged sequentially so obtaining detailed readings along the cross-section can be a tedious process.

22. A. While it is true the micropores and the finest pores are mostly within soil aggregates, it is not true that they hold the water that becomes available for plants. In fact, the water in these fine pores is not available to plants because, due to capillary action, the soil holds on to this water tighter than the plant is able to extract it.

23. C. There is usually no 'unique' relation between the two variables, with the recent wetting and drying history a major variable. Measurement is time-consuming and difficult and this has restricted research into this area. Answers A, B, and D are 'made-up' but may well also be true.

24. C. It's an old concept and somewhat vague, but still commonly used.

25. Answer C is incorrect because water in the vadose zone is usually held loosely. A is one definition of this zone, B and D follow from this definition. E follows from the common restriction of the term 'vadose zone' to mean regions near streams.

26. C. Although A, B and D are conditions under which infiltration can take place, the actual process is matric potential gradients.

27. C. Literally 'rotten rock'. Often minerals from the parent rock are easily extracted from such layers, so they are favoured by miners.

28. B and C. Solutions for flowing fluids are particularly bedevilled by turbulence but real advances have been made in recent years. It is arguable whether such advances will impinge on watershed science. A is untrue in that the equations are not easy to solve. D is untrue in that the equations are still at the heart of the discipline. Newer analytical packages are making solving these equations a more practical operation for the less-mathematically inclined.

29. B. Occasionally piezometers are used to facilitate access to measure other variables. A network of piezometers may allow estimates of the groundwater movement direction.

30. C. The upper layers may approximate E but at some metres depth a water table is usually encountered. D. It's a trick question. You won't usually encounter any aquifers in land-use hydrology. Effects are due to surface runoff.

31. A and B are true. C and D are false. If the piezometer is substantially below the water table a piezometer may differ from the water table; if the intake is above the water table, it won't show anything. A tensiometer usually measures soil tension above the water table.

32. A (this is the definition). B to D are untrue. Pumping from confined aquifers can lead to land settlement but this is not invariable; similarly pumping from unconfined aquifers can also be associated with settlement. Hence E is false.

33. B. Aquitards tend to be lower-permeability layers found in such aquifers. The distinction between aquitards and aquicludes can be vague but the former is usually viewed as being found within a confined aquifer. C to E are just wrong.

34. B only. Transmissivity has the dimensions of $[LT^{-1}][L] = L^2T^{-1}$. Phreatic and confined aquifers show spatial and temporal variation in thickness and conductivity that limit the utility of this parameter. Hence A, C, and D are viewed as generalisations.

35. A and D. A is incorrect and describes a confined aquifer; it is uncommon for these to intersect a land surface and supply water to streams. B is the usual mechanism of stream flow. C is implied by the term 'phreatic aquifer'. Simulations and some field projects suggest that D is incorrect, with groundwater streamlines passing far below the level of the stream.

36. D. Answers A, B, and C may be true sometimes. Ramifying networks of flooded caves can give a complex and spectacular subsurface hydrologic network.

37. C and E. Cave-divers can easily exceed safe depths in such an environment. Groundwater bores and tunnels penetrating confining layers can experience very high pressures.

38. D. Often the groundwater has reached the phreatic surface by a long passage through highly reduced environments. With some exceptions, trees generally do not pass oxygen into the soil.

39. B and E. Although the technology is old, they require constant maintenance if they are to survive. Building quanats has traditionally been a well-paid job and has been handed down from father to son. Unless the work is done carefully there is a high risk of collapse or suffocation. Although used in many parts of the world, it is now easier to use a groundwater bore with a pump in most situations.

40. D. Additionally, the lack of disturbance activities (mining, timber harvesting, recreation activities) provides an extra protection for water quality.

41. A. Forest roads are not typically associated with toxic materials from vehicles because of the low traffic levels. However, urban roads are associated with toxic substances such as hydrocarbons and metals. Usually local materials are used in construction so the presence of contaminants in the construction material is unlikely.

42. A. You need flow data to compute the load.

43. C and (to a lesser extent) E.

44. A. This process is called longitudinal dilution and dispersion. There is a long 'tail' of polluted water but eventually the pollutant passes out of the system.

45. D. Conductivity is highly correlated with ionic content. In many countries Na^+ and Cl^- are the dominant ions so it is commonly used as a measure of salinity. Conductivity does vary with water temperature and many devices attempt to compensate for this.

46. C and E. The others would be viewed as 'intensive measures'.

47. C and D. These are both concentrations (in mg L^{-1}). Thus, multiplication of the measured value by a volume gives the mass in mg.

48. C and to a lesser extent B. BOD is a measure of the amount of oxygen needed to remove waste organic matter by decomposition. Turbidity is often associated with small particles of organic matter from faecal decomposition.

49. B and C are correct. A is incorrect. Use of Secchi disks is not feasible in shallow streams, variable light conditions, and with clear but aerated water. Sometimes turbidity is correlated with other parameters, but the relationship derived tends to be specific for one location only and have a large scatter.

50. B. Because of the purity of many natural streams, instruments 'drift' and different instrument brands give quite different results, although this does not happen in more 'buffered' water. Thus, interstream comparisons can be misleading.

51. C. It would seem that animals tend to move out as humans move in and, for low human populations, not much in the bacterial flora of the stream changes. As populations increase, however, a threshold is passed and increasing population gives increasing bacterial levels.

52. D. To attempt to resolve the issue by chlorinating heavily contaminated water would make the water undrinkable, unusable, and probably toxic.

53. D. The tannin appears similar to weak tea. This can absorb oxygen from the water, which makes survival of oxygen-dependent fauna difficult. The name of a major tributary of the Amazon, the Rio Negro, reflects this behaviour.

54. C. A is acidification. B could be associated with eutrophication in that sediment can carry some nutrients (especially phosphorus) but is not the actual cause, D may cause hypoxia ('dead' water).

55. B. The particles move by both rolling and saltation.

56. C. The correct procedure would be to take enough independent measurements to build up a regression between a newly-installed instruments and the hand-held 'standard instrument', then use this regression for correction. This should be done as standard procedure at the time of each instrument installation. Ideally, the three instruments should give identical readings if they are measuring the same quantity. Our experience was always that it was difficult to make 'instream turbidimeters' agree with laboratory or hand-held instruments because of a range of factors including the shallowness of natural streams, the irregular geometry of their bed, and air-bubbles in the water.

57. A and B, but particularly the latter. To say that it was brilliant intuition belies the amount of dogged and discouraging research he did to make this jump. It was some decades later that scientists showed the mechanism of the link. However, after some resistance there was a worldwide appreciation of the importance of good water supplies and sanitation. This, in turn, led to the search for 'clean watersheds' for cities in the second part of the 19th century – particularly in the relatively 'new' countries of the world.

It would be hard to overstate the importance of this intuitive leap in an age when the science of infection hardly existed.

58. D. Water cannot be oxidised under any naturally occurring situation.

59. B. Water obeys Newton's Law of Viscosity well and is classed as a Newtonian fluid. For hydrology purposes, water is incompressible.

60. B and D are correct. A and C are incorrect – the property is rare and the reason that cans of drink break when frozen is that the molecules are more loosely packed in the frozen state. The ramifications of 'hydrogen bonding' is an area of chemical research.

61. B. Energy includes height energy, pressure energy, and velocity energy. In open channels this usually means the lowest height since pressure is constant and velocity is low.

62. D is incorrect. As ice turns to water its density increases. This means that the ice will float on the water. The chemical behaviour of water is viewed as 'anomalous' with water at low temperatures showing some behaviour akin to crystalline materials.

63. A. Note that the poles have no discernible rainfall.

64. A. This is a 'classic' rainfall mechanism.

65. B and E are incorrect. Answers A, C, and D are correct. B is incorrect because the concept of 'snow ripening' refers to when a snowpack can yield meltwater. This includes warming of the snowpack to 0°C, wetting of the snow, and coarsening of the snow texture. It is considered 'ripe' when it can hold all the water it can against gravity. E is because fresh snow is a porous media and will absorb large amounts of rainfall. This will, of course, facilitate the 'ripening process' but outflow will be delayed. The observation by da Vinci was a generalisation but correct. The observation on snow density comes from Russian literature, but 10% water content is usually used as an average for older snow. Small clearings and clear-cut areas become particularly distinct after snow because of the snow accumulation compared to surrounding forest.

66. B. The method lends itself well to graphic computations.

67. A. B, C and D may be locally evident – there is no 'best answer'.

68. C. This can give some of the highest rainfalls on the earth's surface.

69. D. Hence the name.

70. A. Cloud forests tend to occur in tropical areas with orographic rainfall.

71. C and E are untrue. There is no evidence of cyclicity in droughts. Equally, there is no evidence that droughts are inevitable under climate change. Droughts have been known and feared since humans started paying attention to these things, but to date there is no generally accepted definition of what constitutes a 'drought'.

72. B. 100 mm of rain is equivalent to 1 ML ha^{-1}. Hence 20 mm of rain is 0.2 ML ha^{-1}. It then follows that the volume added to the watershed is 15 x 0.2 ML = 3 ML.

73. D. Such change is usually complex and can be difficult to interpret.

74. C would give the best result. Answer A would be unacceptable because of the random interception loss. B would have too great an interference from the canopy. C would have some interference but meets a commonly accepted standard. D would measure the rainfall in the air and it is hard to know how representative this is of the ground.

75. A. It has proven difficult to make generalisations about the influence of climate change on hydrologic parameters.

76. A. 75%. Roughly 25% is absorbed by the atmosphere (dust, water, ozone) and 50% by the earth's surface (land and sea).

77. A and E are incorrect. Most data analyses suggest that statement A is incorrect – although the data may look to be normally distributed the sequential behaviour of data is not usually that of a normal distribution. Statement E is incorrect – there is no known 'self-correcting mechanism' in annual rainfalls. Statements B, C, and D are borne out by many studies. In particular, Log-Pearson Type III distributions often fit annual data sequences better than a normal distribution. However this does not imply any physical law. The uncertainties expressed in this answer reflects difficulties in 'climate change' research.

78. A. Wood is structurally fixed and trees appear to have no mechanism to reverse hydraulic gradients (although there is conjecture that some can).

79. D. All of A, B, and C. It's a tough life for a tree living in very cold conditions.

80. C. The Bowen Ratio (BR) is used to characterise hydrologic environments (e.g. deserts have a BR often in excess of 10, semi-arid areas have a BR between 2 and 6, temperate forests have a BR of between 0.4 and 0.8, and tropical forests have a BR of between 0.1 and 0.3).

81. A. The effect of atmospheric drivers is mainly dealt with by the vapour pressure deficit and radiation terms.

82. A. High vapour pressure deficit is generally correlated with high rates of transpiration. The other answers could all play a role.

83. B. Some of C and D are captured in the energy balance and mass transfer.

84. B would be the usual place. A tends to be energy limited all the time. C is usually not energy limited. In D it would be rare to have seasonal limitations. E is usually not energy-limited and F would commonly experience water limitations at any time of the year.

85. B. The entropy will manifest itself in randomness and disorder.

86. B. This work has been very fruitful in giving a coherent view of world hydrology and can be viewed as 'top-down' hydrology (in contrast to field measurement approaches).

87. B. The concept of resistance represents the resistance to vapour flow through the soil and plant (xylem and stomata) and to flow upward from the plant, mainly friction.

88. A and B. They are relatively easy to measure. D can be measured with difficulty. C is often inferred by the difference between B and A but can also be directly measured at a plot level.

89. C or D is the concept of 'sealed'. A and B are desirable but reflect the quality of the work, not whether the watershed is 'sealed'.

90. A. All the others are components of the water balance.

91. D is correct. Transpiration includes both the evaporation from the earth's surface and water passed into the atmosphere through plants. A is a component only. B may well be close to it but includes other quantities (deep seepage and error in measurements particularly).

92. E. Perrault demonstrated that the rainfall in its watershed was sufficient to sustain the flow of the River Seine in about 1674. This was a revelation in science at the time.

93. D. Units are usually those of depth. Note that error can go on either side of such an equation.

94. C is true. A is false, the problem is that we don't have a simple method. B is false since we can infer it from differencing rainfall and streamflow. D is sometimes assumed to be true for convenience, but is not always; every opportunity should be taken to check this.

95. D. The effect of thinning is not very long-lasting unless it is continually followed up by more thinning. This gets more difficult as the trees become larger.

96. A and B are correct and involve simple arithmetic (students to check). C is incorrect – it would be grossly optimistic to expect that the entire outflow could be sold to the town. In general, stream water will have many uses and these are sold to users at different prices or given away. A small portion of the water may find its way into an urban supply. Under most laws, water is a 'common good' and not owned by any person. Thus D is incorrect in most societies.

97. A. Streamflow is generated by sub-surface processes and tends to be a steady flow valued by catchment managers. Low evapotranspiration with forests hardly exists, ruling out B and C. There's not enough water for D to occur.

98. B and C are correct. SPAC characterises the energy level in water passing from the soil to the atmosphere. A is false (the water in the leaves is at a lower energy level usually, otherwise there would not be an hydraulic gradient taking water to the top of the trees). D is false (there is no evidence of pumping or pulsing by roots).

99. D. It's a useful concept and often used in calculating aridity and classifying landscapes.

100. B. As a ratio it is dimensionless. Computations generally assume one-sided leaves. In application to specific forests there may be modifications to the definition.

101. C, E, F, and G. Method C is occasionally done but is very laborious. Instrument E requires specialised calibration. F and G are done using well-developed techniques. A porometer measures stomatal resistance of the leaves. B and D are flights of fancy of the authors.

102. A. It is usually complex and multi-factored.

103. C. This is the SPAC (Soil, Plant, Atmosphere, Continuum) concept.

104. C. Carbon is assimilated at higher rates for the same water cost.

105. C. Also called 'Digital Terrain Model (DTM)'. These have become a staple input for some forms of hydrologic modelling. LIDAR has greatly increased the detail available, too.

106. C. The length of record available from 'LandSat' satellites has been immensely valuable in hydrology studies.

107. C. Good ground work is critical to evaluation, interpretation, and successful use of satellite imagery.

108. B, C, and E are correct. Normalised difference vegetation index (NDVI) is not perfect but it is usually a good start in hydrology-based satellite imagery studies. The availability of drones has allowed satellite imagery to be paired with lower-level imagery too.

109. C. Big dams have many values but this is not one. However, it has been shown that once a dam has been there for more than a generation, it tends to become accepted as 'natural' and there is resistance to demolition or restoration.

110. C. It is a very non-random arrangement of components that reduces flow resistance.

111. E is correct. A is the inverse of D (i.e. subtract D from the months of the year and you get A). B would be called annual flooding frequency. C might be computed from D.

112. C and D. Fresh water often limits the economy and must be harvested with care to avoid saltwater intrusion. Look up 'Ghyben-Herzberg lenses' for a description of this type of groundwater behaviour. A would be unusual. B is not the case for a coral island. E is not true.

113. C. While there is some possibility A and B could occur, it is generally believed that preferential trapping is the primary process.

114. A to C are all correct. There was widespread discussion on this point about 50 years ago, which led to many modifications of published statements. It was usually difficult to demonstrate overland flow in forest situations. Because of compaction, agricultural land can have high overland flow.

115. A, B, F, occasionally E. At times of fires streamflow is often zero because of dry conditions.

116. C. However, there are always difficulties with which units to use and how to calculate it. The concept has become less popular in recent years. It probably works best in arid watersheds subject to large, occasional storms.

117. C. The concern is that floods from various tributaries may 'reinforce' one another.

118. C. The structure accommodates a diverse range of biota because of the variety of in-stream habitat. Loss of the pool-riffle structure is a major degradation of the stream often associated with human management. The structure is usually consistent within a given landscape.

119. A, B and D are likely to be correct. By understanding how the system, in particular soils and vegetation, has developed we can better understand how the system may respond to stressors. C is possible, but usually higher production infers more intensive management.

120. C. From the Latin *saltus* for leap, apparently. Can also occur in aeolian transport. Sediment particles move downstream in discontinuous 'jumps'.

121. A. Usually it is a combination of many factors that combine but the rain exacerbates any weaknesses and the weight of the water dramatically increases the slope loading. Earthquakes can also generate many landslides but are less common.

122. D. Floods from A and C are common events. An example of B is the Vaiont Dam disaster. An example of E is the Mount St Helens disaster. Wait long enough and D will probably happen somewhere.

123. B. Photographs of rainbows often include the shadow of the photographer pointed towards the centre.

124. B. The term 'optimum' implies a minimum extinction possibility.

125. Mainly B. This may have been stored in the slopes for years and is pushed out by the pressure of incoming rainfall infiltration. This is replaced by 'new water'.

126. A, D, and E. The concept is that early in the storm the runoff is from areas close to the stream. As the storm continues this 'source area' expands.

127. A and C are incorrect. The Aswan Dam has been controversial and has delivered a wide range of both benefits and disbenefits. The flow from the Nile is no less important now than in ages past. Nilometers go back to Pharaonic times and have given some of the longest hydrographic records available. The origins of the

River Nile were a major source of debate in Europe in the mid-19th century, with funded research expeditions to find this source.

128. B and D. This has been a pivotal piece of modern hydrology with the suggestion that we tend to get long 'runs' of good years followed by bad years. It used a very long sequence of data from the flooding of the Nile River in Egypt.

129. C. The idea is attributed to English Franciscan friar William of Ockham (1287–1347).

130. A. It's one approach to dealing with the complexities of nature.

131. D. The models always seem to take vastly longer than envisaged to get working.

132. A and C are true. D, and E may sometimes be true, B is false.

133. B. 'Topmodel' routed both surface and subsurface water from one grid cell to another based on a topographic index.

134. A is incorrect (in science, a 'double-blind' experiment with replication would occupy that hierarchy). B is correct. C is 'Maybe' – the inherent variability of watersheds means that a true replication of Land-use 1 probably can't be obtained. D is correct. E is incorrect – any watershed is on a trajectory of change over time, so the control will also change. F is correct in that the calibration will allow gross differences in behaviour of the basins to be characterised.

135. C and E. Often the skewness of distributions of storms is so great that analysts hypothesise that large storms come from a separate distribution to that of small storms.

136. A and D are correct. B is incorrect in that, for that combination of units, K would be 10^4. (To convert A from hectares to m^2, multiply by 10^4. Remembering that 1 mm depth of water on 1 m^2 is 1 litre, then multiplying by 10^4 gives units of litres per hour.) C is incorrect in that although parameter C is a function of I, a large value of I implies a high C coefficient.

137. D. Usually some of the points would be labelled to show time. A change in the gradient indicates a change in the relationship between the two (e.g. deforestation). The method subordinates detail (by integration) to give an overview. It was a favourite form of graphical analysis in pre-computer days.

138. D. Although all of these could contribute to high flows, we would really need data such as land use, vegetation type, and soil properties.

139. A. The effects can be very severe.

140. A. Base flow is viewed as a smooth, continuous outflow formed by groundwater processes contributing to stream channels.

141. C. 0.9. There is about a 10% loss due to wetting of surfaces, capillary storage, and evaporation.

142. A, C, E, and F. There are usually many compromises inherent in such strategies.

143. D. In urban hydrology this leads to a range of infiltration structures and 'trapping dams'.

144. B. Traffic continually mobilises sediment on the wet road surface, adding to the proclivity for pollution.

145. A. This assumes the crop had the physical structure to intercept the water and hold it in the root zone. Other hydrologic characteristics would differ, however.

146. A, B, and C would be accepted. Because the experiment may go on for 50 years or more, the concept of a 'stationary control' becomes doubtful. Answers D and E come under the category of 'wishful thinking'. Because each watershed is unique there is little possibility of true 'replication', and usually the cost and area involved prohibit this. A 'double-blind' approach may be possible but to date the logistics and the analytical difficulties have precluded this.

147. D. The response is typical of this situation. Thinning (A) would give a somewhat similar response but the magnitude of the increase would be less.

148. D would be the usual response. The flow increase is usually associated with site preparation killing off the pasture.

149. B is the likely response. As the pine ages its water use increases but it never matches the water use of the mature native forest.

150. C. This experiment has not often been done but the results are consistent and suggest that the trees along the streams do a disproportionate share of the watershed's transpiration.

151. B and C are relatively easy to show. We've never heard of anyone finding A, D, or E in upland watersheds.

152. D would be very likely, followed by B and A. Causes C and E would be most unlikely in usual circumstances. Turbulence-induced level fluctuations in measurement structures is a major issue in hydrology measurement.

153. A, C, and D. The streamflow at the start of the storm is 'antecedent flow' and is an indicator of wet watersheds. The work showed that the wetter the watershed the greater the watershed response to a given period of rainfall. B is a figment of imagination (but may well be true, since this indicates a porous mantle). E is a figment of the imagination.

154. C. This reflects the diminution of evapotranspiration for a period, followed by the growth of understorey and canopy trees.

155. All of the above. In 1967 Alden Hibbert formulated three long-lived hypotheses: (1) reduction of forest cover increases water yield; (2) establishment of forest cover on sparsely vegetated land decreases water yield; and (3) response to treatment is highly variable and for the most part unpredictable. Examination of many studies suggest that the variation observed is covered by these hypotheses (particularly the third). The presence of snow in the areas particularly appears to increase the difficulty of making a correct prediction.

156. B. Liquid water cannot flow below this layer, so that permafrost environments tend to be very boggy in summer.

157. B. It's a failure that has been repeated many times around the world since.

158. B. Most hydrologic change are cumulative effects, but scientific hydrology finds it difficult to deal with more than one or two effects at a time.

159. B and perhaps D. This was the sub-title of a conference on cumulative effects some years ago. In general, attempts to show the existence of cumulative effects in complex watersheds have not been markedly successful and have often been controversial.

160. B. The Wagon Wheel Gap (1910–1926) was the first such experiment and looked at the impact of deforestation. The Emmental paired watershed measurement project started in 1903 and continues today but did not have an experimental component and hence was not an 'experiment'. Coweeta, Croppers Creek, Mokobulaan, (and many others) followed the model of the early watershed experiments.

161. B and D. He was one of the first to question the role of 'overland flow' in forest environments, and did much experimental work.

2 Matching Questions

PART 1. Match the Label with the Illustration Place-holders

1. The matched hydrograph-hyetograph in Fig. 2.1 shows stream-flow in two adjoining streams, together with rainfall intensity during a major spring-time (in Australia) storm. Match the names with the appropriate labels.

    1. Hydrographs (rising)

    2. Hydrographs (recession)

    3. Hyetograph

    4. Time (day of month)

    5. Stream flow, Ls^{-1}

    6. Rainfall intensity, mm hour^{-1}

    7. Period of maximum 1-hour rainfall intensity

    8. Peak flow for Ella Creek

    9. Antecedent flow for Clem Creek

© Leon Bren and Patrick Lane 2021. *Key Questions in Hydrology and Watershed Management* (L. Bren and P. Lane)
DOI: 10.1079/9781789249682.0002

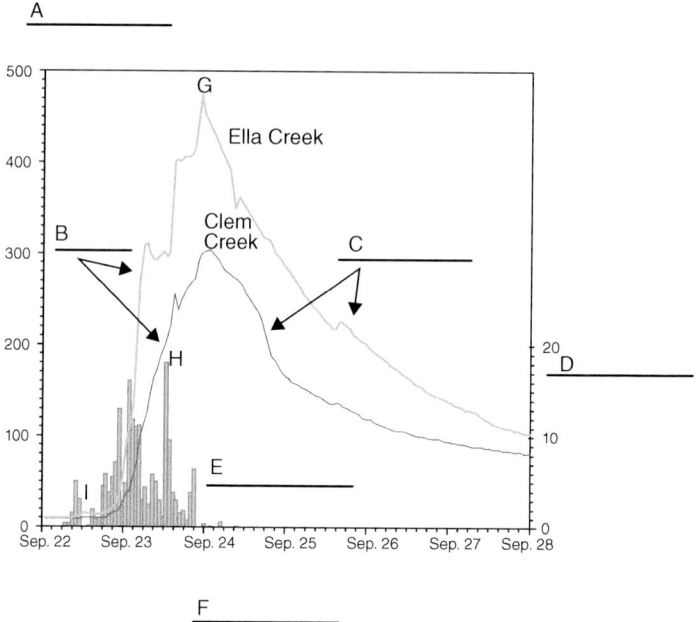

Fig. 2.1. Illustration courtesy of Springer Nature.

2. **In the photographs of the V-notch weir, allocate the names to the labels on the illustration.**

   1. V-notch

   2. Data-logger (recorder) housing

   3. 'Trash-rack'

   4. Measuring well

   5. Stilling pond

   6. Nappe showing *Vena contracta* (two locations)

   7. Weir water level (base of notch at zero)

   8. Sharp-edged blade

Fig. 2.2.

3. The illustration in Fig. 2.3 shows an oblique aerial photograph of a research watershed in south-eastern Australia. In this the slopes have been just cleared of native eucalypt forest and replanted with radiata pine. The streamflow and water quality was measured for some years before this transition and for a growth cycle of the pine. Match the names below with the placeholders on the illustration.

1. Treatment area

2. *Pinus radiata* on slopes

3. Watershed (ridge)

4. Riparian zone

5. Springhead and convergent zone

6. Outflow measurement weir

7. Native eucalypt forest

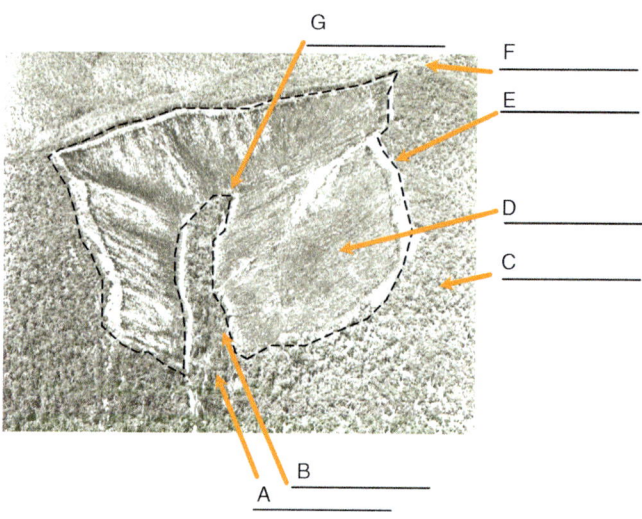

Fig. 2.3. Illustration courtesy of Springer Nature.

PART 2. Fill in the Gaps Using the Words Provided

4. **Insert the most appropriate word from the box below in the descriptive text.**

> A. downslope, B. rainfall, C. headwater, D. convergent, E. colluvium

About small streams. 'Life is not easy for a ___(1)___ stream to survive. Firstly, it must have enough watershed area and the correct ___(2)___ geometry to provide a groundwater outflow for its springhead. Secondly, it must have enough continuing ___(3)___ to sustain the water flow. Thus, in the major 1997–2009 drought in southern Australia, 'permanent' streams were anything but that because of inadequate rainfall. Thirdly, the streamflow must be great enough to remove the accumulation of ___(4)___ moving ___(5)___ into the stream.'

Write the correct letter next to the number 1–5.

1.   4.

2.   5.

3.

5. **Insert the most appropriate word from the box below in the descriptive text.**

    > A. hydrostatic, B. drained, C. contaminated, D. phreatic, E. pumping

    Hydrology and tunnels. In building tunnels, one faces the option of either draining the tunnel or sealing it to prevent groundwater access. If the tunnel is ___(1)___ it can put a very heavy and costly ___(2)___ load on the tunnel management and lower ___(3)___ surfaces for many kilometres around. Further, heavily ___(4)___ groundwater has to be disposed of in an environmentally responsible manner. Sealed tunnels avoid this but can be subject to extremely high ___(5)___ pressures which require massive walls and can endanger lives if a sudden failure occurs.

    Write the correct letter next to the number 1–5.

    1.   4.

    2.   5.

    3.

6. **Insert the most appropriate word from the box below in the descriptive text.**

    > A. minimal, B. scarce, C. quality, D. low-energy, E. potable, F. porous, G. recovery

    Aquifer storage. Aquifer recharge is sometimes cited as an excellent way of rehabilitating and storing water for later use. It is viewed as a low-cost ___(1)___ water supply option. There is ___(2)___ water loss by evaporation. Passage through ___(3)___ media may remove small particles and organisms. However, the water will not stay where it is placed, it may suffer deterioration in water ___(4)___, and ___(5)___ may be difficult. It is hard to find good examples. Suitable aquifers may be ___(6)___. Sometimes the injection water will contaminate reasonably ___(7)___ water in the aquifer.

    Write the correct letter next to the number 1–7.

    1.   5.

    2.   6.

    3.   7.

    4.

7. **Peter Eagleson died in 2021. He wrote a leading text *Dynamic Hydrology* in 1975 which placed scientific hydrology on a much more rigorous footing. The text below is extracted from the citation at the presentation to him of the 'Horton Medal' by Ignacio Rodriguez-Iturbe for excellence in research. Select the appropriate word from the box and insert in the paragraph.**

A. rainfall networks, B. dynamic, C. stochastic, D. flood, E. fluid-mechanics, F. modeling

Peter Eagleson's research. 'Here was somebody who was tackling the ____(1)____ of overland flow with the scientific standards of ____(2)____. At the same time, finally somebody had arrived in the field who, using techniques from signal analysis, produced criteria for the optimum density of ____(3)____. I did not know Pete at that moment, but I remember reading many times his paper on the optimum density of rainfall networks and feeling the excitement of the author's original mind at work. It is the same feeling one has when reading his 1972 classic paper on the dynamics of flood frequency, where for the first time the ____(4)____ nature of ____(5)____ is analytically tied to the physical-dynamic characteristics of the basin response. This intimate linkage between the probabilistic aspects of the phenomena, so crucial to hydrology, and the ____(6)____ modeling of the physical aspects of the processes has become the trademark of Peter Eagleson's research and has had a profound influence in the field.'

Write the correct letter next to the number 1–6.

1.   5.

2.   6.

3.

4.

8. **The text box contains a list of instruments or techniques which you might find a hydrologist (or, at least, a researcher) using.**

A. Neutron probe	F. Pluviometer
B. Capacitance probe	G. Tensiometer
C. Nephelometer	H. Evaporation pan
D. Manometer	I. Data logger
E. Infiltrometer	J. Conductivity meter

For each sentence, select the instrument(s) that might be used.

1. A _____ (mounted on the weir wall) is often used to check that the recorded level matches the actual water level in the notch of a V-notch weir.

2. A _____ is sometimes used to measure surface levels of soil moisture, while a _____ is used to measure deeper levels.

3. A _____ is used in the laboratory to measure turbidity in water samples.

4. A _____ is often used to measure a surrogate for salt levels in natural waters.

5. Measurement of soil water potentials in wet soils can sometimes be made with a _____.

6. A _____ is an alternative name for what is commonly called a rain gauge.

7. The concept of an _____ is that water is applied to a bounded area of soil and the volumetric rate at which it passes into the soil is measured.

8. The _____ loss is sometimes multiplied by 0.7 to give a rough estimate of evapotranspiration.

9. Most streamflow measurement weirs use a _____ to record the height of water passing over the weir crest.

Write the correct letter(s) next to the number 1–9.

1.	6.
2.	7.
3.	8.
4.	9.
5.	

9. **The text box below contains a list of flow-measuring devices or associated concepts. Place the correct one in the sentences below.**

A. current meter	F. sharp-crested
B. Cipoletti	G. broad-crested
C. 45° V-notch weir	H. flume measurement devices
D. laminar	I. 'rated-section'
E. acoustic device	J. compound weir

1. The _____ gives excellent measurement at low flows but does not have much capacity.

2. A _____ notch which makes a transition from one shape to another gives a continuous but difficult-to-determine weir rating.

3. A common older form of _____ was like a hand-held water turbine with a revolution counter. Modern ones often use an _____ and have no moving parts.

4. _____ usually require less stream gradient than a weir and hence can be easier to install.

5. _____ weirs are easy to install and have a good flow capacity but are of low accuracy.

6. A _____ weir has a flat, level bottom and sloping sides. They are often a good compromise between accuracy and capacity.

7. A_____ weir usually involves the water passing over an edge formed by sharpened steel. This gives very reproducible flow properties.

8. A _____ is where flow as a function of water level has been determined for a natural reach of stream. Accuracy is inadequate for higher-level scientific work.

9. Ideally, on a sharp-crested weir, the flow nappe should spring clear of the crest with a smooth, _____ flow.

Write the correct letter(s) next to the number 1–9.

1.	6.
2.	7.
3.	8.
4.	9.
5.	

10. **The text box below refers to four water contamination issues which have achieved worldwide coverage.**

A. arsenic	F. *Giardia* and *Cryptosporidium*
B. arsenic-related	G. lead
C. *Escherichia coli*	H. watershed
D. naturally	I. groundwater
E. lower-income	J. agricultural

Select items from the text box and place in the appropriate gaps.

1. The Flint water crisis occurred in 2014 when a city in Michigan, USA had its water supply changed. The new water leached _____ from ageing pipes, causing illness in residents. It was argued that this disproportionately affected _____ residents. In 2021 charges were laid against managers involved.

2. The Walkerton _____ outbreak occurred in 2000 in Canada. It was due to poor chlorination monitoring and a water supply which pumped _____ from an aquifer with faecal contamination. More than 2000 people became ill and six people died. An inquiry showed poor procedures across the spectrum of water supply.

3. Well water contaminated by _____ in Bangladesh is considered one of the most devastating public health crises in the world. Almost a quarter of the country's population, an estimated 39 million people, drink water _____ contaminated by this element, which can silently attack a person's organs over years or decades, leading to cancers, cardiovascular disease, developmental and cognitive problems in children, and death. An estimated 43,000 people die each year from _____ illness in Bangladesh.

4. The Australian city of Sydney was thought to have the pathogens _____ in its major water supply dam in 1998. The origin was thought to be poor _____ land management in the dam _____. The crisis showed many deficiencies in both land and water management and led to a complete reorganisation of the water supply agency.

Write the correct letters in order next to the number 1–4.

1. (2 letters)
2. (2 letters)
3. (3 letters)
4. (3 letters)

11. **The text box below contains a list of rivers or wetlands that have some claim to fame in hydrologic circles.**

A. Murray	F. Rio Grande
B. Jordan	G. Yangtze
C. Colorado	H. Rio Negro
D. Sundarbans	I. Brahmaputra
E. Nile	

Select items from the text box and place in the appropriate gap.

1. The _____ River supplies the south-western USA and Mexico, but is famous for the fights over the allocation of its water.

2. The River _____ in Australia sustains 'Australia's food bowl' but has inadequate water to meet all the irrigation, urban water supply, and conservation requirements of the four Australian states through which it flows.

3. The _____ River was the cause of much puzzlement in 19th-century scientific circles as to where it came from.

4. The _____ River has one of the world's largest dams and the largest hydropower project ever built.

5. The _____ is a large tributary of the Amazon River and is famous for its black water associated with 'humic acid' derived from decomposing vegetation.

6. The _____ provides a natural border between the USA and Mexico and ownership of water rights has been an intensely political question.

7. The Ganges River and the _____ River merge to form one of the world's great freshwater wetlands – the _____.

8. The _____ River holds major significance in Judaism and Christianity since the Bible says that the Israelites crossed it to reach the Promised Land and that Christ was baptised in it.

Write the correct letter(s) next to the number 1–8.

1.   5.

2.   6.

3.   7. (2 letters)

4.   8.

12. **Famous Freshwater Wetlands. The text box below contains names of some famous wetlands or wetland issues. Match these with the missing words in the text.**

A. salt water	G. Kissimmee River
B. marsh Arabs	H. swamp cypress
C. international	I. regulation
D. evaporated	J. *Eucalyptus camaldulensis*
E. Iran and Iraq	K. River Murray
F. diversions	L. Okavango River

Match up the word in the text box with the brief wetland description below.

1. Barmah Forest, Australia is a very large expanse of almost a single species – red gum (_____) – and exists in an intimate relationship with the _____. The forests are under threat from river regulation limiting necessary forest flooding.

2. Okavango Delta (Botswana) is a swampy inland delta formed at the terminal end of the _____. All water is ultimately _____ or transpired. The area is famous for its biotic diversity.

3. Everglades (Florida, USA) is a large wetland formed by discharge of the _____. It is famed for its prolific biota but under threat from human encroachment and water _____.

4. Atchafalya Basin (Louisiana, USA). This is the largest wetland in the USA, and is an alternative course of the Mississippi River to the ocean. Famed for its biota including _____ (*Taxodium distichum*). The wetlands are under threat from human encroachment, flow modification, and dredging of ocean channels allowing _____ penetration.

5. Euphrates/Tigris Wetlands, between _____, are the largest wetland of Western Eurasia. This area is sometimes thought to be the home of the human species. The area supports a diverse biota and a unique human culture ('_____'). It is under threat from drainage, warfare, and human encroachment.

6. The RAMSAR Convention on wetlands aims to halt the word-wide loss of wetlands and to conserve those which remain. It is viewed as the first modern, _____ conservation treaty.

7. The word 'Ochoco' is a First-Nation word for willows. The Ochoco forests of the USA are one of the world's great floodplain forest types and suffer from adverse effects of river _____.

Write the correct letter(s) next to the number 1–7

1. (2 letters)	5. (2 letters)
2. (2 letters)	6.
3. (2 letters)	7.
4. (2 letters)	

PART 3. Match the Left-hand Side and the Right-hand Side

13. Hydrograph Features. Pair the name in Column A with the definition in Column B.

Column A	Column B
A. Storm Hydrograph	1. The maximum flow that can be associated with a discrete period of rainfall.
B. Diurnal Variation	2. A continuing decline in streamflow after a wet period.
C. Recession	3. An estimate of the flow hydrograph that would have occurred had there not been a rainstorm.
D. Peak Flow	4. A repeating cycle of flow variation with a 24-hour period, superimposed on other flow hydrograph events.
E. Base Flow	5. The streamflow variation associated with a period of rainfall.

Write the correct letter next to the number 1–5.

1.	4.
2.	5.
3.	

14. Paleohydrology is interpretation of past hydrology episodes. The table below shows common remnants of paleohydrology in landscapes and their usual interpretation. Match the list of indicators on the left with the most likely cause from the right-hand column.

Indicators	Causes
Larger Streams	
A. Flood levels painted on buildings	1. Deposition from old floods
B. Oxbow lakes	2. Colluvium-erosion sequence and climate change
C. Arcuate sand dunes in flat areas	3. Bank flood deposits from old floods
D. Impact scars in tree ring sequence	4. Old floods causing tree trunk damage
E. Layered sediment deposits	5. Reduced stream energy, climate change
F. Natural levee banks	6. Paleo-lake site
G. Bank erosion and entrenchment	7. Paleo-stream sites
Headwater Streams	
H. Buried stream beds	8. Removal of flow obstacles, woody debris
I. Alluvial fans on bedrock	9. Ancient floods viewed as worthy of note

Write the correct letter next to the numbers 1–9.

1.   6.

2.   7.

3.   8.

4.   9.

5.

15. Places Commonly Cited in Watershed Courses

Match the place name in Column A with why it is often discussed in hydrologic literature in Column B.

Column A	Column B
A. Melbourne, Australia	1. Site of many paired watershed experiments
B. Atchafalya Basin, USA	2. 'Bull Run' closed watershed policy
C. Amazon River	3. Controversial diversion of water for Los Angeles
D. Columbia River, USA	4. Early, continuing scientific hydrology study
E. Portland, USA	5. Closed watershed policy in older watersheds
F. Aswan Dam	6. River crossing many international boundaries
G. Emmenthal, Switzerland	7. Early very large dam associated with huge benefits/disbenefits
H. Owens River, USA	8. Trade-off of power generation and fish values
I. Coweeta, USA	9. Wetland on alternative Mississippi River course
J. Danube River, Europe	10. One of the biggest watersheds in the world

Write the correct letter next to the numbers 1–10

1.

2.

3.

4.

5.

6.

7.

8.

9.

10.

16. General Watershed Science Topics

Pair the named item in Column A with the brief definition in Column B.

Column A	Column B
A. *Am/Be* source	1. High, short-lived flows after forest fires
B. Dupuit Approximation	2. Steady-state equation of groundwater flow
C. Bedrock	3. Plot of one time integral against another
D. Arsenic	4. Decomposing time record into frequencies
E. Fourier analysis	5. Sediment material deposited by slowing water
F. Antecedent flow	6. Simplifies groundwater problems
G. Laplace Equation	7. Impervious watershed base
H. Double-mass plot	8. Natural and common groundwater contaminant
I. Spike flows	9. Source of neutrons for a neutron moisture probe
J. Alluvium	10. Stream flow before a big storm event

Write the correct letter next to the numbers 1–10.

1.   6.

2.   7.

3.   8.

4.   9.

5.   10.

17. Hydrologic Processes

Pair the process in Column A with what it is in Column B.

Column A	Column B
A. Fog drip	1. First-order stream
B. Transpiration	2. Groundwater emerging at a point on a hillside
C. Overland flow	3. Flow which would have occurred if no rain
D. Headwater	4. Rainwater passing through the forest canopy
E. Throughfall	5. Water captured by watershed from fog
F. Storm flow	6. Water running across the watershed surface
G. Spring flow	7. Streamflow associated with rainfall
H. Base flow	8. Water passed by plants to atmosphere
I. Stemflow	9. Vague term for water passing downslope
J. Interflow	10. Water running down tree stems

Write the correct letter next to the numbers 1–10.

1.   6.

2.   7.

3.   8.

4.   9.

5.   10.

18. Equations and What They Estimate

Pair the name of the equation in Column A with what it estimates in Column B.

Column A	Column B
A. Manning's Equation	1. Early equation for rate of infiltration
B. Boussinesq Equation	2. Evapotranspiration rates
C. Penman Equation	3. Flow in a channel
D. Darcy's Law	4. Rate of groundwater movement
E. Penman–Monteith Equation	5. Evaporation rates from water
F. Dupuit Approximation	6. Unsteady state behaviour of groundwater
G. Bernoulli's Equation	7. Rate of flow through a soil column
H. Green–Ampt Equation	8. Energy balance of flowing water
I. Phillip's Equation	9. Conservation of mass in flowing streams
J. Continuity Equation	10. Later equation for rate of infiltration

Write the correct letter next to the numbers 1–10.

1.   6.

2.   7.

3.   8.

4.   9.

5.   10.

19. Measurements and Instruments

Pair the measurement in Column A with the appropriate instrument(s) in Column B.

Column A	Column B
A. Area on a map	1. Nephelometer
B. Flow velocity	2. Recording pluviometer
C. Deep soil moisture level	3. Hook gauge, staff, or manometer
D. Sample turbidity	4. Capacitance probe
E. Weir water level	5. Planimeter (electronic or mechanical)
F. Sap velocity	6. Conductivity meter (often)
G. River stage	7. Heat pulse measuring set
H. Shallow soil moisture	8. Staff gauge
I. Rainfall intensity	9. Neutron probe
J. Sample salinity	10. Current meter

Write the correct letter next to the numbers 1–10

1.   6.

2.   7.

3.   8.

4.   9.

5.   10.

20. Common Surrogates

Column A contains measurements often used as convenient surrogates for the more difficult or expensive measurements of Column B in watershed science. Match Column A and Column B.

Column A	Column B
A. Conductivity	1. Stream flow
B. Leaf area index	2. Transpiration rate
C. Soil moisture	3. Salinity
D. Turbidity	4. Transpiration capacity
E. Periodic pan evaporation	5. Soil water potential
F. Stream height	6. Periodic evapotranspiration
G. Snow depth	7. Sediment and faecal pollution
H. Sap flow velocity	8. Water-dissolved carbon content
I. Fish-eye crown photographs	9. Snowpack water content
J. Water colour	10. Leaf area index

Write the correct letter next to the numbers 1–10

1.   6.

2.   7.

3.   8.

4.   9.

5.   10.

21. Scientists who made a big contribution to hydrology

Pair the named scientist in Column A with their contribution in Column B.

Column A	Column B
A. John Hewlett (1922–2004)	1. Soil potential theory, 'Pi' dimensionless groups
B. Daniel Bernoulli (1700–1782)	2. Concepts of contours, perhaps watersheds
C. Mikhail Budyko (1920–2001)	3. Variable source area model
D. Edgar Buckingham (1867–1940)	4. Equation for evaporation from water
E. Robert Horton (1875–1945)	5. Linking rainfall, evaporation, and energy input
F. Joseph Boussinesq (1842–1929)	6. Equation partitioning energy in flow
G. Howard Penman (1909–1984)	7. Terminology for ordering streams
H. Arthur Cayley (1821–1895)	8. Useable equation of groundwater flow
I. Newell Strahler (1918–2002)	9. Concept of overland flow on agricultural land
J. Jules Dupuit (1804–1866)	10. Simplification of groundwater theory

Write the correct number next to the letters A–J

A.   F.

B.   G.

C.   H.

D.   I.

E.   J.

22. All about transpiration

Match each item in Column A with the correct statement in Column B.

Column A	Column B
A. Xylem	1. Openings in the leaf through which gas and water are exchanged
B. Stomata	2. A measure of how easily water can move out of a leaf
C. Guard cells	3. Cells by which water travels upward within the plant
D. Root hairs	4. Cells that swell and shrink to control stomatal aperture
E. Stomatal conductance	5. Part of the plant that absorbs water from the soil
F. Turgor	6. Leaf pressure too low to support tissue
G. Wilting	7. Layer of air in direct contact with leaf
H. Leaf boundary layer	8. Adequate leaf pressure to sustain transpiration

Write the correct number next to the letters A–H

A.   E.

B.   F.

C.   G.

D.   H.

Answers

PART 1. Match the Label with the Illustration Place-holders

	Question No.	
1.	**2.**	**3.**
1.B	1.H	1.E
2.C	2.A	2.D
3.E	3.D	3.F
4.F	4.B	4.A
5.A	5.C	5.G
6.D	6.E,I	6.A
7.H	7.G	7.C
8.G	8.F	
9.I		

PART 2. Fill in the Gaps Using the Words Provided

				Question No.				
4.	**5.**	**6.**	**7.**	**8.**	**9.**	**10.**	**11.**	**12.**
1.C	1.B	1.D	1.F	1.D	1.C	1.G,E	1.C	1.J,K
2.D	2.E	2.A	2.E	2.B,A	2.J	2.C,I	2.A	2.L,D
3.B	3.D	3.F	3.A	3.C	3.A,E	3.A,D,B	3.E	3.G,F
4.E	4.C	4.C	4.C	4.J	4.H	4.F,J,H	4.G	4.H,A
5.A	5.A	5.G	5.D	5.G	5.G		5.H	5.E,B
		6.B	6.B	6.F	6.B		6.F	6.C
		7.E		7.E	7.F		7.I,D	7.I
				8.H	8.I		8.B	
				9.I	9.D			

PART 3. Match the Left-hand Side and the Right-hand Side

Question No.									
13.	**14.**	**15.**	**16.**	**17.**	**18.**	**19.**	**20.**	**21.**	**22.**
1.D	1.E	1.I	1.I	1.D	1.H	1.D	1.F	1.D	1.B
2.C	2.H	2.E	2.G	2.G	2.E	2.I	2.H	2.H	2.E
3.E	3.F	3.H	3.H	3.H	3.A	3.E	3.A	3.A	3.A
4.B	4.D	4.G	4.E	4.E	4.F	4.H	4.B	4.G	4.C
5.A	5.I	5.A	5.J	5.A	5.C	5.A	5.C	5.C	5.D
	6.C	6.J	6.B	6.C	6.B	6.J	6.E	6.B	6.G
	7.B	7.F	7.C	7.F	7.D	7.F	7.D	7.I	7.H
	8.G	8.D	8.D	8.B	8.G	8.G	8.J	8.F	8.F
	9.A	9.B	9.A	9.J	9.J	9.C	9.G	9.E	
		10.C	10.F	10.I	10.I	10.B	10.I	10.I	10.J	

3 Fill-in-the-blank Questions

About Water

1. Water is a _____ fluid because it obeys the law of viscosity.

2. Water is viewed as having a negative tensile strength of around 250 _____. This may be easily exceeded in subsurface water movement in ____ conditions. There are many difficulties in _____ of this.

3. The specific gravity of pure, air-free water at 20°C decreases from 1.002 to 0.923 as pressure goes from 1 kg cm^{-2} to 2000 kg cm^{-2}. Thus water is not _____, but this is still a useful assumption for the range of _____ encountered in hydrology.

4. The velocity of sound in water at 10°C is about 1445 m sec^{-1}. In general, a fluid is considered incompressible if the velocity of flow is small compared with the speed of _____. This is clearly achieved in _____.

5. Because water is such a good conductor of _____, it follows that fish live in a richer _____ environment than we air dwellers.

6. Aerated water is a mixture of air bubbles and _____, and is typically white in colour. The presence of air gives it a low density, and this makes _____ in aerated water a dangerous proposition because it may not have enough buoyancy to support a human.

7. Although water has a complex _____ diagram showing the state (solid, liquid, _____) as a function of temperature and pressure, the water behaviour in the zone of earth-surface conditions is a simple ice-water-vapour _____ with increasing temperatures.

© Leon Bren and Patrick Lane 2021. *Key Questions in Hydrology and Watershed Management* (L. Bren and P. Lane)
DOI: 10.1079/9781789249682.0003

8. When the ratio of observed water vapour pressure to the _____ water vapour pressure at a given temperature is expressed as a percentage, it is called the relative _____.

9. In three dimensions, Bernoulli's Equation is meant to apply to a 'stream ____' (i.e. bounded by stream-lines). This implies no _____.

10. The Reynolds Number is a dimensionless quantity which characterises the _____ in the channel flow. Because it is dimensionless, it has no _____.

11. The Froude Number is a _____ quantity which characterises the speed of a disturbance in water to the velocity of the water. It is a measure of how much the water is 'leaning' on water _____ of it. If it is 1 then the water is more or less about to pass over a waterfall.

12. As a stream of water increases speed, its cross-sectional area must correspondingly _____. This is known as the principle of continuity.

13. In general, the _____ of snow and rough water is due to bubbles of ____ entrained in the medium.

14. When water dissipates energy by turbulence, about 1% of the energy is emitted as _____. The noise spectrum of this is close to a '_____ noise', with no dominant frequency present.

Snow Hydrology

15. If a column of snow is taken and melted, the resultant depth of water in the sampling tube is called the ____ _____ _____.

16. Graupel is where super-cooled _____ freezes directly on falling snowflakes, forming 2–5 mm balls of rime. In the _____ language it means sleet.

17. Slab _____ are when an entire layer of ____ slides off the slope to create the avalanche. They are viewed as the most dangerous type and are often triggered by victims.

18. Snow is very sensitive to the ____ at which it is loaded or stressed and this leads to slab _____.

19. Wind can _____ snow ten times faster than snow falling from _____.

20. An average snow density of 0.10 gm cm^{-3} is commonly used for _____-
_____ computations, but Soviet studies have shown _____ variations due to wind from 0.06 gm cm^{-3} to 0.34 gm cm^{-3}.

21. Before appreciable _____ can occur from snowpack the greater
part of the snow must be brought to the melting point and then given
the latent heat of _____.

22. Sublimation of _____ is where water passes directly from ice (or snow)
to _____ without becoming water.

23. The conditions at high altitude in the Himalaya that favour (snow) sublimation are low atmospheric _____, high wind speed, and dry ___.

24. Advection melt of snowpack refers to the _____ of heat from rainfall into the snowpack, leading to _____ of the snow.

25. Conduction melt of snow refers to the upward flow of _____ from the
ground surface into the _____ _____. This often produces melt at the
soil–snow interface.

26. The albedo of a surface is the ratio of reflected to incident light. _____
has an unusually high albedo and burnt surfaces have an unusually
____ albedo.

27. As snow ages, its albedo generally _____.

28. Clathrate hydrates are crystalline, water-based _____ resembling ice in
which non-polar _____ are trapped within the polar ice structure.
These occur in large quantities in polar environments.

Units and Dimensions

29. Hydrologic computations are most easily done in __ units. The most
common mistakes relate to the number of zeroes. Non-metric units
like _____ and _____ can and are used but require a greater amount
of practice to achieve competence.

Hydrographic Measurement

30. After much trial and error, about 100 years ago, the circular ___ _____
was decided upon as the most efficient measuring device. This is of a
standard size (203 mm) and has a _____ rim literally to split raindrops. In
the USA about one-billionth of the rainfall is measured in gauges.

31. Hydrography is the science or art of measuring _____ variables. It started to develop as a technology about the middle of the 19th century. It was rapidly taken up by the countries being newly settled such as Australia and Canada. As a result some of these countries have some of the longest records of _____ in the world.

About Precipitation and Rainfall Processes

32. The term 'precipitation' in hydrology refers to the input of water; this is generally as ____ or ____. Other forms (e.g. hail) exist but are of minor importance.

33. Long sequences of rainfall data tend to be _____ in nature. This tends to obscure long-term trends.

34. As moist air is forced upwards by a mountain range it usually cools and _____ results. This is called _____ precipitation. After passing the mountain range, the air may descend and warm due to the increased pressure. This results in a _____ effect that is very drying.

35. 'Fog' is an _____ of tiny water droplets, too small to fall.

36. A droplet is called a '____ _____' when it is between 0.5 and 4 mm. Their diameter can be measured by briefly exposing a tray of light oil to rainfall and measuring the drops under a _____.

37. The size of __ ____ is limited by air resistance. If they become too big this breaks them up as they fall. If the air is moving _____ then larger drops than usual may impinge on the earth's surface.

38. The 'dew-point' is the _____ at which cooled air becomes saturated with water. The closer this is to the actual temperature, the more _____ it feels.

39. When air is swept up, it may be cooled to or below its dew-point. In these circumstances ___ may form, but this often requires particles of ice or dust on which the raindrop condenses. If these are ___ present, the air may become super-saturated.

40. The big, spectacular _____ associated with _____ are called 'cumulonimbus', literally meaning 'heaped rainstorm'.

41. A set of depth-area-duration curves can be compiled to characterise _____ at any point on the earth's location. In general, as the area increases the depth of rainfall decreases for a given storm _____.

42. 'Foehns', 'Chinooks', and 'Santa Ana' winds are all examples of hot dry winds resulting from rain-shadow air _____ on the ____ side of mountains.

43. On average, atmospheric _____ are observed to be at their hottest at the earth's_____. The presence of trees can have a distinct cooling effect due to _____.

44. The dry adiabatic lapse rate is the rate at which the temperature of a dry air parcel changes in response to the _____ or expansion associated with elevation change (and assuming no ____ exchange with surroundings). It is about -9.8°C km^{-1}.

Watershed and Water Balance Equations

45. The water balance of a watershed involves totalling, on one hand, the input _____ and on the other the outputs (_____, _____) and the change in watershed water storage in units of depth. The difference between the two aggregated quantities is a crude measure of _____.

46. Transpiring water is held in a tree at a _____ pressure relative to that of free water at the same height.

47. 'Sap flow velocity' in trees is measured by injecting _____ into the sapwood and then timing the passage of this pulse between points.

48. 'Fog drip' is usually a _____ component of the hydrologic cycle and is due to the _____ of fog on the foliage of high-altitude forests. This then drips to the ground or passes down the trunk as ____ ____.

49. The Australian forest tree called mountain-ash (*Eucalyptus regnan*) is known as a species in which the water use of the forests varies as a function of the ____ of the forest.

50. Thinning of forests reduces the density of trees. This usually results in a _____-_____ increase in streamflow. After _____ years the tree crowns fill the gap and the increase in yield disappears.

51. In principle an estimate of evaporation can be made using an _____ balance approach. In practice this is difficult.

52. In general, plant _____ open in the day and close at night. Two exceptions are cactus and pineapple for which the stomata open only at _____.

Plant Evapotranspiration and SPAC

53. If a tree (or forest) is called a 'phreatophyte' it means that it can gain water by passing roots _____ or _____ the water table.

54. A lysimeter is a device used to measure the amount of _____ of growing plants (crops or trees).

55. In _____ lysimeters, the amount of water in the measurement chamber is assessed by continued weighing. Because the weights of the soil pots are substantial, and the water loss may be small, the weighing needs to be _____.

56. Ground-surface evaporation may cause an upward _____ potential gradient sufficient to overcome _____.

57. Consumptive plant use of water is sometimes defined as annual _____ plus water stored within the plant tissues.

58. In tree stems, _____ is the hydraulic conductor of low resistance.

59. Potential evapotranspiration is the _____ which could occur if the water supply was unlimited both to the soil surface and the stomata, and the _____ were open.

60. The _____ equation has achieved iconic status in hydrology for its representation of the relationship between annual evapotranspiration and long-term-average water and energy balance at _____ scales. This has led to new approaches to hydrology science.

Soil Moisture and Vadose Zone

61. A macropore is a void in the soil such that capillary forces play little or no _____ in the movement of _____.

62. Colluvium refers to loose, unconsolidated sediment deposited at the base of hill slopes by rain-wash, sheet-wash, slow continuous _____ creep, or a combination of those processes. If a stream does not have enough energy to wash colluvium downstream it becomes buried and is often called a _____.

63. The term vadose zone is sometimes used to describe the very wet area near streams where _____ water will reach the surface.

64. A common assumption in watershed hydrology is that the underlying native ___ ____ is impermeable. This is rarely tested and usually found not to be the case because of _____ networks.

65. The Navier–Stokes equations are sometimes viewed as a theoretical basis for the behaviour of _____. These can be expressed as a series of partial _____ equations and are not always simple to _____.

66. The characterisation of soil pores as small, tortuous _____ tubes and the use of such analogues in developing theory is now viewed as fanciful and unrealistic.

67. A good example of unsaturated flow is the phenomenon of damp in _____ houses. In this the moisture _____ between wet soil and the dry house interior is enough to 'pull' water through seemingly impervious materials such as bricks.

68. Soil moisture is sometimes measured by a neutron probe. This requires an _____ _____ to be inserted into the soil to allow the probe to be lowered to the correct depth.

69. Water above the water table is said to be held by the soil under _____.

70. When water travels in a soil in which at least some of the pores are air-filled and in contact with free air, the soil is said to be _____.

71. When a forest soil is strongly heated, it may develop a water _____ in which a drop of water on the soil surface sits as a spherical drop and does not infiltrate into the soil. This lack of infiltration can lead to major _____ events after fire.

72. 'The critical surface tension method' of measuring water repellency of forest soils involves placing drops of increasing _____ concentration on the soil until infiltration occurs within a given time. This is used in assessing _____ impacts.

73. The mass wetness of a soil is the _____ of water relative to the mass of dry soil particles. This is also known as the '_____ water content'. In contrast, the 'volumetric water content' is that computed as the _____ of water per unit volume of soil.

74. Ideally, soil should be viewed as a 'three-_____ system'. The phases are soil, water, and _____.

75. In forests, 'soil' is often a rather vague term since it may also be viewed as _____ or oxidised rock and may lack the ordered horizons of agricultural soil.

Groundwater

76. Much of groundwater hydrology and soil physics is based on assumptions of a continuum. This implies a 'minimum representative _____' below which the assumption of a _____ is no longer valid.

77. An influent stream passes water _____ to groundwater. An effluent stream _____ water from groundwater.

78. In situations where aquifers are compressible, removal of groundwater may lead to surface _____.

79. The famous Laplace Equation of groundwater flow is a _____-_____ solution. Hence it cannot have a maximum or minimum value of groundwater head except at a domain _____.

80. Unsteady state solutions of groundwater require a solution domain, one or more governing _____, a set of _____ conditions governing what is happening at the edge of the solution domain, and a set of _____ conditions saying what is the state at the start of the solution.

81. The concept of an unsteady-state partial differential _____, as found in groundwater theory, requires a solution of change in one, two, or even three directions over _____.

82. Outflow of an aquifer to a stream or drain is generally assumed to occur at a _____ ___.

83. Before water from watershed slopes can pass into a stream, it must achieve the _____ of free water. This means that effectively only water from the saturated zone can pass from the slopes into the _____.

84. The Ogallala Aquifer is a shallow _____-_____ aquifer surrounded by sand, silt, clay and gravel and located beneath the Great Plains in the USA. It is one of the world's largest aquifers but has, for years, suffered from declining _____ _____. These have been associated with over-_____ for agricultural purposes.

85. Mound springs (e.g. Dalhousie Springs) are oases in arid Central Australia. Over millennia, _____ deposits have caused the springs to rise above surrounding plains in elevation, giving a small lake on a hill. These support a diverse fauna. Declining _____ pressures in the Great Australian Artesian Basin are putting these at risk.

86. The Great Artesian Basin is located in Australia and is the largest and deepest artesian _____ in the world. By definition an _____ basin has a phreatic pressure great enough to bring water to the surface but overuse has caused declining _____on recent years.

Satellite Imagery and Remote Sensing Hydrology

87. The consensus of _____ is usually that the largest changes in the field in years will be caused by the advent of new satellites allowing measurement on a scale hitherto impossible.

Water Quality

88. When a 'slug' of pollutant is introduced to a stream, the body of pollutant moves downstream and attenuates. This is referred to as _____ dilution and _____.

89. *Escherichia coli* is a water-borne _____ that is often used as an indicator of _____ pollution.

90. When taking water quality samples, finding a sampling location where stream turbulence overcomes _____ within the cross-section is often an important criterion.

91. There is commonly a _____ correlation between _____ measures of water quality such as sediment concentration and turbidity, but the relationship differs from stream to stream.

92. Many soils show thixotropic properties when mixed with water. This means that the _____ decreases once shaken or stressed. Once this perturbing force is removed the viscosity increases. This can be a major factor in earthquake-induced _____ and many other hydrologic environments. Many river deposits are thixotropic, making them dangerous to perturb.

Modelling and Quantitative Hydrology

93. Double mass plots are a form of graphical hydrologic analysis in which the _____ outflow of a 'treated' watershed is plotted as a function of the _____ outflow of the control. Statisticians dislike this form of analysis because it does not give a clear indication of _____.

94. MIKE SHE, HEC, HMS, and TUFLOW are examples of commercially used _____ _____. They all have a big learning overhead.

95. The 'Nash and Sutcliffe' Coefficient of Efficiency is sometimes used to test the fit of models. A perfect fit gets a score of _. A score of _ means that it is no better than an average. A poor fit of the model gets a large _____ score.

96. In hydrologic modelling, use of a sequence of measured data to make estimates of parameters which allow the model to reproduce the data is known as _____. Use of a second, independent sequence of data to test whether the model provides a good fit to data is called _____.

97. A '_____-___' model is one in which an empirical model (e.g. regression) is used to estimate the required property. There is no pretence that the physical reality of the subject is being _____,

Urban Hydrology

98. Urban watershed runoff increases _____ loads to receiving waters because of greater sources of pollutants and hydraulically _____ connections of the sources to the drainage network.

99. 'Water-sensitive urban design' projects aim at returning _____ characteristics of the watershed closer to those of a _____ watershed and reducing the _____ carried to the streams.

100. Manning's Equation of flow is often used in urban runoff computations. It works reasonably well in these because the equation's hydraulic radius _____ is definable for constructed channels (unlike natural streams). The '_____ coefficient', however, often varies with the depth of streamflow.

101. When the vegetation in a drainage channel becomes submerged the 'Manning's n' coefficient of _____ drops because the vegetation lays over in the direction of flow and no longer offers the same _____.

102. Using infiltration as a way of managing urban _____ has the benefits of making a more natural watershed and sustaining _____ values, but can carry risks of potential damage to building foundations and mobilisation of 'legacy' _____.

103. Reducing the impact of urban stormwater _____ on receiving waters can be difficult. Strategies include stormwater _____ to reduce peak flows and restore baseflows, bio-retention systems (rain-gardens and _____). Unfortunately, some 'armouring' of receiving channels may still be necessary to maintain _____ stability.

General Hydrology

104. In overland flow, water passes _____ the watershed surface along the most downward path to the stream.

105. Hydrology is the study of water's passage across ____ _____ _____.

106. Parts of Brazil, New Guinea, and Bangladesh are distinguished by having extensive areas with very high _____ ___. These can be 3,000 mm pa or more.

107. The 'variable source area model' has it that storm hydrographs are generated by rainfall landing on the _____ parts of the landscape near the stream.

108. In watershed data, if the measured streamflow (and other liquid outflows) is subtracted from the measured rainfall (and other liquid watershed inflows), the net difference is a measure of _____.

109. A periodic variation superficially like a sinusoid with a 24-hour period is called a '_____'.

110. The valuation of water emanating from a forest may involve computation of impacts over long periods of time. Often the concept of compound interest is used to bring costs and benefits to the same point in ____.

111. Interception is the loss of water in a watershed water balance due to rainfall wetting _____ and being re-evaporated.

112. The _____ of a clear sky is thought to result from a _____ of short wavelengths (violet) of the sun's white light by air molecules.

113. _____ of water in flowing stream channels helps biota retain their position in the moving water stream. This is often associated with rocks and large woody _____.

114. In general, the mathematical formulations derived from hydraulics theory in a laboratory are not applicable in headwater stream hydrology because of the extreme _____ of the surfaces encountered and the corresponding _____ in flows.

115. Terms such as '_____ floods' and '___ weather' are pejorative and reflect human values since they are rarely harmful from a hydrologic or natural viewpoint.

116. Perhaps the oldest recorded _____ observer was Ecclesiastes (*c.* 300 BC) who wrote 'All streams flow into the _____ , yet the sea is never full. To the place the streams come from, there they return again' (Ecclesiastes 1:7).

Land-use Hydrology

117. In a paired watershed experiment, the distinguishing feature is that one of the watersheds acts as a _____. This has no _____ other than routine management during the course of the experiment.

118. A stream _____ is a set-back of any economic use from the stream frontage. Typically these are viewed as protecting water quality and restricting stream temperature variations.

119. Lahar is an Indonesian term for a _____ phenomenon describing a hot or cold mixture of water, ash, and rock fragments that flows down the slope of a _____ and enters a river valley. They look like a rolling slurry of wet concrete and can be amongst the most dangerous effects of volcanic activity. The mix is _____ and can set like concrete once it stops flowing.

120. That doyen of American forest hydrologists, John Hewlett, is reputed to have quipped at a forest _____ conference 'Ahh, hydrograph analysis, the last respite of the desperate _____'. I think of this every time I decide to go down that path.

Forest Fires

121. When a forested watershed is burnt, the hydrograph commonly features _____ _____ in which short-lived but very high flows occur. These can be responsible for massive _____.

122. Burning of a forest is often a major initiator of slope erosion events including _____ flows, _____, and _____ fan formation where the flow gradient decreases.

123. The advent of 'megafires' in the 21st century means that, in fire-prone areas, the ____-_____ designs of large dams needs to be re-evaluated. In particular, fire-affected watersheds may develop _____-flows of a magnitude hitherto not experienced.

Answers

1. Newtonian.
2. Atmospheres, dry, measurement.
3. Incompressible, pressures.
4. Sound, hydrology.
5. Sound, water.
6. Water, swimming.
7. Phase, vapour, transition.
8. Maximum, humidity.
9. Tube, turbulence.
10. Turbulence, units.
11. Dimensionless, downstream.
12. Decrease.
13. Whiteness, air.
14. Noise, white.
15. Snow water equivalent.
16. Rainfall, German,
17. Avalanches, snow.
18. Rate, avalanches.
19. Deposit, storms.
20. Water-equivalent, density.
21. Runoff, fusion.
22. Snow, vapour.
23. Pressure, air.
24. Transfer, melting.
25. Heat, snowpack.
26. Snow, low.
27. Decreases
28. Solids, molecules.

29. SI, feet, pounds.
30. Rain gauge, sharp.
31. Hydrology, streamflow.
32. Rain, snow.
33. Stochastic.
34. Rainfall, orographic, rain-shadow.
35. Aerosol (or colloid, or emulsion).
36. Raindrop, microscope.
37. Raindrops, downwards.
38. Temperature, humid.
39. Rain, not.
40. Clouds, thunderstorms.
41. Rainfall, duration.
42. Descending, lee.
43. Temperatures, surfaces, evapotranspiration.
44. Compression, heat.
45. Precipitation, (streamflow, evaporation), error.
46. Negative.
47. Heat.
48. Minor, condensation, stemflow.
49. Age.
50. Short-lived, several.
51. Energy.
52. Stomata, night.
53. Into or below.
54. Evapotranspiration.
55. Weighing, accurate.
56. Capillary, gravity.
57. Evapotranspiration.

58. Xylem.

59. Evapotranspiration, stomata.

60. Budyko, watershed.

61. Role, (or influence, effect), water.

62. Downslope, drainage line.

63. Capillary or ground.

64. Bedrock, fracture.

65. Water, differential, solve.

66. Capillary.

67. Brick, gradient or differential

68. Access tube.

69. Tension.

70. Unsaturated.

71. Repellency, erosion.

72. Ethanol, fire.

73. Mass, gravimetric, volume.

74. Phase, air.

75. Weathered.

76. Volume, continuum.

77. Down, receives.

78. Subsidence.

79. Steady-state, boundary.

80. Equations, boundary, initial.

81. Equation, time.

82. Seepage face.

83. Potential, stream.

84. Water-table, water levels (or phreatic pressures), extraction.

85. Evaporite, groundwater or phreatic.

86. Basin, artesian, pressures.

87. Hydrologists.

88. Longitudinal, dispersion.

89. Pathogen or bacteria, faecal.

90. Stratification.

91. Positive, conservative.

92. Viscosity or shear-strength, landslips.

93. Cumulative (or accumulated or summed), cumulative (or accumulated or summed), errors.

94. Hydrologic models.

95. 1, 0, negative.

96. Calibration, verification.

97. Black-box, modelled.

98. Pollutant, efficient.

99. Hydrologic, forested, pollutants.

100. Parameter, roughness.

101. Roughness, resistance.

102. Stormwater, ecological, pollutants.

103. Runoff, infiltration, wetlands, channel.

104. Along, across, or over.

105. The earth's surface.

106. Annual rainfalls.

107. Saturated.

108. Watershed evaporation or evapotranspiration.

109. Diurnal variation.

110. Time.

111. The vegetated surface.

112. Blue, scattering.

113. Turbulence, debris.

114. Roughness, turbulence.

115. Disastrous, bad.

116. Hydrologic, sea.

117. Control, disturbance or treatment.

118. Buffer.

119. Worldwide, volcano, thixotropic.

120. Hydrology, hydrologist.

121. Spike or flashy flows, erosion.

122. Debris, gullying, alluvial.

123. Peak-flow, peak.

4 True or False Questions

Read each statement below and indicate in the corresponding brackets whether it is [True] or [False].

1.	SI units commonly used in watershed hydrology include centimetre (cm) for distance, hours for time, and gram (g) for weight.	[...........]
2.	Bernoulli's Equation can be applied quite well to streamflow channels.	[...........]
3.	The phreatic boundary of a watershed always coincides with the topographic boundary.	[...........]
4.	Water always flows in the path that maximises its rate of shedding energy.	[...........]
5.	Turbidity is a measure of water quality made by shining a beam of light through water and measuring the ratio of scattered to transmitted light.	[...........]
6.	If the measured turbidity is multiplied by the volume of water sampled, one obtains an estimate of the amount of matter in the water.	[...........]
7.	Completely pure water has a low turbidity because of scattering of light by water molecules.	[...........]
8.	Studies of water quality in natural watersheds are often rendered difficult by the purity of the natural waters.	[...........]
9.	In many mathematical models of watershed processes the watershed soil is assumed to be anisotropic and homogeneous in its properties. These are good assumptions.	[...........]
10.	Water in a stream emits a small percentage of energy dissipated as noise. This is an atonal 'white noise'.	[...........]

© Leon Bren and Patrick Lane 2021. *Key Questions in Hydrology and Watershed Management* (L. Bren and P. Lane)
DOI: 10.1079/9781789249682.0004

11.	Most energy emitted by water serves to heat up the water. However, because of the high heat-storage capacity of water, it is almost impossible to detect any temperature rise in a natural system.	[............]
12.	In a smoothly flowing stream, when water passes over an obstacle on the stream bed, there is a small rise in the surface water level.	[............]
13.	The 'head' of water is defined as the energy per unit weight. The dimension of this quantity is that of length.	[............]
14.	Unsaturated soil physics often uses analogues of Darcy's Law but in which the hydraulic conductivity, K, varies as a function of the negative hydraulic head. As the soil dries, the hydraulic conductivity increases because there are more pathways for the water to go through.	[............]
15.	Use of soil-physical formations in forested watersheds is often unsatisfactory because water will pass along macropores in the soil without passing through smaller soil pores.	[............]
16.	In water quality, forming a 'rating curve' showing the relation between the water quality parameter and the flow rate can be complexed by hysteresis depending on whether the flow is rising or falling.	[............]
17.	Interception refers to rainfall that wets the forest canopy and then is re-evaporated. It is such a minor component that it is not worth quantifying.	[............]
18.	A 'runoff curve' is a graph showing the amount of streamflow ('runoff') as a function of annual rainfall (with occasional other variables as well).	[............]
19.	To obtain an accurate measurement of rainfall in a forest, a rain gauge should have at least a 45° cone of clearance and, ideally, a 30° (from the horizontal) cone of clearance.	[............]
20.	Rainfall elasticity is defined as the percentage change in streamflow for a 1% change in rainfall. For most watersheds this would be less than 1%.	[............]
21.	Streamflow hydrographs are a mixture of cyclic, non-cyclic, and occasional random variations. The usual natural cyclic variations are the annual streamflow cycle and the daily (diurnal) cycle.	[............]
22.	A knick-point in a stream is something that confers stability on the bed of the stream and allows it to resist the shear (and other) forces of the water passing along the stream channel. Logs and woody debris form common knick-points in streams in forested landscapes.	[............]

23.	Darcy's Law is that the flow rate through a soil sample is proportional to the hydraulic gradient across the sample, with the constant of proportionality being called the hydraulic conductivity. This could be described as a one parameter model.	[..........]
24.	Groundwater is held so deep in the catchment slopes and moves so slowly that it cannot play a role in headwater stream hydrology.	[..........]
25.	Streams emanating from karst landscapes may have very different subterranean watersheds than their topographic watershed.	[..........]
26.	The maximum hourly rainfall intensity achieved during a rainfall event is a good indicator of the peak flow achieved by a stream.	[..........]
27.	A graph showing rainfall intensity as a function of time is known as a hydrograph.	[..........]
28.	In general, the annual rainfall of a year is independent of the annual rainfall of preceding years.	[..........]
29.	Weir ratings are unaffected by the temperature of the water passing though the weir.	[..........]
30.	In a paired watershed project, one should have at least 30 years calibration between the 'control' and the watershed to be treated if accurate results are to be obtained.	[..........]
31.	When you assemble a year of daily streamflow data and plot out its distribution, you will get something quite close to a normal distribution of flow.	[..........]
32.	Streams emanating from forested catchments have high-quality water that is low in bacterial levels.	[..........]
33.	Statistical comparisons of 'treatment effects' involving water quality are usually rendered difficult or impossible because of non-compliance with the requirements of statistical sampling.	[..........]
34.	When stream sediment is measured in a stream from a forested watershed during a storm event, the sediment levels rise during the period of increasing streamflow and then hold constant over the duration of the event.	[..........]
35.	Stream beds are reasonably approximated as random collections of stream sediment and debris dropped as flows decreased during previous hydrologic events.	[..........]
36.	Streams tend to deposit sediment on the outside of bends. This makes them 'self-correcting' as the channel tends to straighten out from the resultant deposition.	[..........]

37.	Large pieces of woody debris embedded in forest streams are indicators of poor forest management.	[..........]
38.	An amazing property of water is that it expands as it freezes. This has great ecological consequences in cold climates.	[..........]
39.	Erosion of the landscape by running water generally reflects poor land management.	[..........]
40.	That famous historic figure, Leonardo da Vinci, made many trenchant observations about the hydrologic cycle in his notebook. He might be regarded as an early hydrology scientist.	[..........]
41.	Ablation refers to the transition from water vapour to snow or ice. It is an important process in maintaining snowpacks.	[..........]
42.	As snow ages, the water content in a snowpack decreases because the water is frozen into ice.	[..........]
43.	Because of the presence of wind, openings in a forest (e.g. clear-cuts) have less snow accumulation in them than the more sheltered adjacent forests.	[..........]
44.	Reynolds Number is a dimensionless combination of parameters that indicates whether flow in a stream of water is laminar or turbulent. Virtually all streams in watersheds exhibit laminar flow.	[..........]
45.	Two identical rain gauges located at the same level and side by side should achieve rainfall readings within 1% of each other.	[..........]
46.	A reference book says 'the maximum depth from which water can be returned to the surface either by plants or by capillarity provides the lower boundary of what is called 'soil' by agronomists.' This is also a good definition for use in forested watersheds.	[..........]
47.	Headwater streams are often both receiving water from groundwater and passing water to groundwater at different points along their length.	[..........]
48.	A downslope movement of water parallel to the watershed surface through macropores sometimes occurs and is called 'interflow'.	[..........]
49.	The movement of infiltrated water vertically downwards in a watershed soil is correctly called 'percolation'.	[..........]
50.	The phreatic surface in a watershed is the network of points where the pressure of water in the soil is equal to that of a free water surface at the same height.	[..........]
51.	Two piezometers are side by side. The deeper one has a higher piezometric head than the shallower one. This means that the groundwater is moving downwards.	[..........]

52.	A large groundwater bore that is used for pumping can also be used to monitor groundwater levels.	[...........]
53.	In Strahler ordering, when a fourth-order stream meets a fourth-order stream, the order of the new stream is fourth-order.	[...........]
54.	If 120 mm of rain falls on a 46 ha catchment, and 50% passes into a reservoir, then the volume of augmentation of the reservoir is 27.6 ML.	[...........]
55.	A stream has a cross-sectional area of 2.5 m². A current meter indicates a net velocity of 2.5 km hour⁻¹. The stream flow rate is 6.25 m³ hour⁻¹.	[...........]
56.	If a stream falls by 2 m in 1000 m, then the hydraulic gradient is 0.02 = 2%.	[...........]
57.	A smoothly flowing stream passes from a broad, flat section to a swiftly flowing section. In section 1 the area of cross-section and the water mean velocity are 12 m² and 0.5 ms⁻¹, respectively. In section 2 the velocity is 1.5 ms⁻¹. It follows that the area of cross-section would be one-third of that of section 1.	[...........]
58.	In a smoothly flowing stream, a dip in the surface of the water followed by a rise indicates a hole in the stream bed.	[...........]
59.	A 'ring around the moon' is associated with refraction caused by ice crystals hanging in a particular orientation in the upper atmosphere.	[...........]
60.	To observe a natural rainbow, one must always have one's back to the sun.	[...........]
61.	As raindrops become smaller, the colours created in a rainbow become less intense and more whitish.	[...........]
62.	'Continuity' in a water streamflow means that kinetic energy is conserved at any two cross-sections.	[...........]
63.	A small dam designed to reduce flow velocity and control soil erosion by localised deposition is known as a 'check dam'.	[...........]
64.	Removal of sediment from a natural stream by passing it into a settling pond is an excellent conservation strategy.	[...........]
65.	A 'meadow' is a vague term that is sometimes used as a synonym for 'wetlands'.	[...........]
66.	Wetlands need to be kept under water permanently if they are to thrive.	[...........]
67.	Wetlands are associated with floodplain deposition of eroded materials.	[...........]

68.	The Froude Number is defined as $\dfrac{v}{\sqrt{gd}}$ where v is the velocity, g is the gravitational constant, and d is the depth of flow. It has the dimensions of [L].	[...........]
69.	When water moves through unsaturated soil, it is usually the result of a pressure gradient 'pushing' the water through.	[...........]
70.	Model optimisation usually involves modifying a subset of two or three parameters so that the model reproduces some 'calibration data'. This maintains the physical reality inherent in the model.	[...........]
71.	The common water bio-contaminants *Giardia* and *Cryptosporidum* are easily removed by chlorination of water.	[...........]
72.	In Yellowstone Park, there has been a documented link between the hydrologic regime, growth of willows, and herbivory by beavers and elks.	[...........]
73.	Large dams often benefit downstream forests by preventing flooding and stabilising flows.	[...........]
74.	Although the willow and poplar forests of the Central USA occur commonly along rivers and on flood plains, they are not a flood-dependent species.	[...........]
75.	Most flooding species can be categorised by a 'hydroperiod' of flooding. If they do not get this they will collapse.	[...........]
76.	Flooding species are usually viewed as responding to the water inputs. However, this can equally be explained by response to the nutrients carried by the water.	[...........]
77.	In flooding areas, the greatest diversity in habitats is created by large floods.	[...........]
78.	'Riparian forests' and 'flooding forests' are synonyms.	[...........]
79.	Trees found in flooding forests often have specialised features to cope with long floods.	[...........]
80.	River regulation usually deprives flooding forests of water.	[...........]
81.	In the swamp cypress forests of the lower Mississippi, a threat to the freshwater forests is the penetration of salt water due to modification of the bayous.	[...........]
82.	The Australian river red gum (*Eucalyptus camaldulensis*) is widely planted around the world because it can tolerate both drought and flooding very well.	[...........]
83.	The term 'fen', 'marsh', and 'bog' are often synonyms for non-forested wetlands.	[...........]

84.	'Precipitation excess', 'groundwater excess', and 'floodwater excess' wetlands is a classification of wetlands that is useful and easily used.	[............]
85.	'Lacustrine' is defined as 'pertaining to lakes'.	[............]
86.	A water level of 2 m imposes a pressure of about 19.6 kPa on the stream bed.	[............]
87.	The hydraulic radius of a channel is defined as the ratio of cross-sectional area to wetted perimeter of a water conduit. It is a useful parameter to measure for most headwater streams.	[............]
88.	Discharge equations based on the stream slope (e.g. Manning's Equation, Chezy Equation) can be made to work, but usually require a lot of practice with measurement to give reliable results.	[............]
89.	The 'bottom of a watershed' is the bedrock a few metres under the saprolite zone.	[............]
90.	Small samples of bedrock are taken and measurements on these show that the rock samples are impermeable. It is then, a reasonable assumption to assume that the bedrock makes little contribution to the hydrology.	[............]
91.	Most of the earth's groundwater is held in sedimentary rocks.	[............]
92.	Once you know the grain-size distribution of sedimentary rock, estimating the permeability using one of a number of 'universal equations' is straightforward.	[............]
93.	A 'Ramsar Wetland' requires that the wetland be managed in a special and unusual way to enhance wetland values.	[............]
94.	A cut-off meander is formed when flood-prevention levees cut a section of a river off to keep the levee straight.	[............]
95.	'Resacas' in Mexico, 'Billabongs' in Australia and 'Oxbow Lakes' in the USA are all local names for cut-off meanders.	[............]
96.	A levee bank is an earthen wall that stops a river passing onto its adjacent flood plain.	[............]
97.	In hydrologic terms, a levee bank is one of the simplest and most effective things one can do to give flood protection.	[............]
98.	For stability, a levee bank wall should have a vertical drop of 1.5 units to a horizontal run of 1 unit.	[............]
99.	The ability of levee banks to carry heavy trucks on top is important for their stability because it guarantees maintenance during flood periods.	[............]
100.	The role of the levee banks around New Orleans in the 1927 Mississippi River Flood are famous in flood-hydrology history.	[............]

101.	Cut-off meanders are important for river ecosystems because they often provide habitat refugia for species that cannot survive in the main channel.	[..........]
102.	Wetlands have many sources of variation but tidal movements are not one of them.	[..........]
103.	In large wetlands, the presence of upstream erosion may be a threat to their existence.	[..........]
104.	Floods are best viewed as hydrologic disasters.	[..........]
105.	'Deflation' of organic soils is associated with reduction of the organic compounds in the soil under severe flooding.	[..........]
106.	Deflation of organic soils can increase the probability of flooding in low-lying areas.	[..........]
107.	A levee built on one side of a river dramatically increases the probability of flooding on the other side.	[..........]
108.	If groundwater pumping can lead to surface deflation, it follows that pressurised groundwater injection into the aquifer can reverse the process.	[..........]
109.	After the French Revolution, France showed adverse hydrologic effects of land use that were not experienced in the USA for another century.	[..........]
110.	Land abandonment in the high mountain catchments of Europe is leading to forest expansion, and this is reducing stream flows in these parts.	[..........]
111.	Studies have shown that, after the American Civil War, stream yields over much of the American south diminished.	[..........]
112.	In general, after establishment and crown closure, water use by growing trees is independent of tree age. But one eucalypt species – *Eucalyptus regnans* – in Australia does show a changing water use with tree age.	[..........]
113.	A large landslide falling into a large lake can lead to catastrophic flooding both upstream and downstream.	[..........]
114.	Bedrock on watershed slopes is impermeable and plays no role in stormflow hydrology.	[..........]
115.	In streamflow measurement, most recorders measure the height of water above a datum and use a measured relation between water level and volumetric flow rate.	[..........]
116.	In streamflow measurements, the water level is usually recorded in a well hydraulically connected but physically remote from the water source. This is to reduce the impact of temperature fluctuations in the stream water.	[..........]

117.	A data logger implies use of a digital signal. Hence the old-fashioned paper chart recorder sometimes encountered in gauging stations, is not a data logger.	[...........]
118.	The concept of a 'control watershed' in a long-term, paired watershed study involving forests implies that the control remains unchanged. This is easily achieved in such studies by setting aside one watershed.	[...........]
119.	'New water' can easily be created by burning oxygen and hydrogen. This is often the basis of rocket fuels. This water is an addition to the world's reservoir of water.	[...........]
120.	The word 'water' appears to be very, very old but variants of it can be found in many European languages.	[...........]
121.	The origin of the earth's water is unknown.	[...........]
122.	Water (as a liquid) is common on the planets of our solar system and in space generally.	[...........]
123.	Water (as a compound) occurs in the earth's magmatic core.	[...........]
124.	As far as we know, the earth is distinctive amongst planets because of the visible erosion trails associated with the passage of water across its surface.	[...........]
125.	Erosion by water on the earth's surface is a bad thing and should be stopped wherever possible.	[...........]
126.	Adhesion refers to the ability of water to stay as a drop and cohesion refers to the ability of water to stick to surfaces.	[...........]
127.	Water has about the properties one might expect from studying liquids derived from common elements.	[...........]
128.	Water can dissolve more substances than any other liquid.	[...........]
129.	'Virtual water' of a good is a measure (albeit imperfect) of how much water was used in creating that good.	[...........]
130.	A kilogram of meat has a virtual water content of, say, 700 kg of water. This means that to grow that 1 kg of meat, irrigation and transpiration of the grass consumed 700 kg of water.	[...........]
131.	A piece of meat has a virtual water content of 700 kg of water. Should you choose not to eat that meat, the world would have an additional 700 kg of water.	[...........]
132.	The photosynthetic efficiency of plants and the virtual water content of plant material are both measures of the amount of water used per unit of mass in producing plant material. They are, then, the same thing if measured carefully.	[...........]
133.	Excluding Antarctica, the driest place (least rainfall) on the earth's surface is the Atacama Desert in Northern Chile.	[...........]

134.	Drought is sometimes defined as 'an acute water shortage'. This is an adequate definition for hydrology studies.	[..........]
135.	The concept of drought is difficult to apply to areas that are naturally very dry.	[..........]
136.	Irrigation with large, multi-linked dams is well known as an economic solution to drought for many countries.	[..........]
137.	Centre pivot irrigation always gives a large, green circle on a satellite image.	[..........]
138.	Archaeologists in the Tigris-Euphrates River system (now Assyria, Sumad, and Akkad) can find successive civilisations built on irrigation which then collapsed from salinity. These spanned millennia.	[..........]
139.	A common and excellent irrigation drainage strategy is to collect saline drainage water and pass it into a major river, allowing dilution to reduce concentrations.	[..........]
140.	In 1983 the diversion of irrigation salt waste from the Central Valley of California into a salt marsh near San Francisco became an affair which was followed internationally.	[..........]
141.	'Potable' is a synonym for drinkable when describing water.	[..........]
142.	Crystal-clear water from mountain streams generally has negligible bacterial and other microscopic counts.	[..........]
143.	In managed watersheds, we can rank domestic animals in the order sheep, goats, pigs, and cattle in the decreasing order of the likelihood of passing a disease that affects humans into water.	[..........]
144.	Wild pigs are often particularly effective transmitters of water-borne diseases in municipal watersheds around the world.	[..........]
145.	'Three pipe' water systems are sometimes suggested for new towns. The three pipes carry fresh water, gas, and sewerage.	[..........]
146.	A hydrographer specialises in the measurement of streamflows and sampling of streams for water quality.	[..........]
147.	The basic premise of a stream rating curve is that the stream stage (aka water level) is a stable indicator of both the cross-sectional area and the mean flow velocity.	[..........]
148.	In general, at a stream gauging station in a long, smooth channel reach, the geometry of the downstream channel does not affect the stream rating.	[..........]
149.	The terms 'vadose zone' and the 'unsaturated zone' are usually synonyms.	[..........]
150.	A 'perched water table' is a zone of saturated soil within an unsaturated zone.	[..........]
151.	About 3 years after clearfall logging, there is a period of enhanced slope instability if the slopes are unstable.	[..........]

152.	On unstable hill slopes, development of forests can be both a protector and a cause of landslips.	[..........]
153.	Streamflow and traffic flow are quite analogous and can be modelled using similar mathematical approaches.	[..........]
154.	'Once one knows the geological type of the bedrock one can make some guesses about the watershed hydrology.'	[..........]
155.	Hydraulic conductivity and aquifer porosity have a stochastic variation at any location. This should be allowed for in computations.	[..........]
156.	Canopy conductance is leaf area multiplied by leaf conductance.	[..........]
157.	An area has a high ratio of potential evaporation to precipitation. This an energy-limited system.	[..........]
158.	In forested landscapes, the water entering the stream network during a storm event is most likely to be the rainwater from that event.	[..........]
159.	Soil erosivity is a measure of the susceptibility of the soil to erosion.	[..........]
160.	Altitude (or elevation) may be analogous to increasing aridity.	[..........]
161.	Latent heat flux can represent evaporation.	[..........]
162.	Identify true or false for some of the typical pollutants in urban runoff.	
	(a) Suspended solids	[..........]
	(b) High levels of mercury	[..........]
	(c) Silver from photographic processing plants	[..........]
	(d) Nutrients (particularly nitrogen and phosphorus)	[..........]
	(e) Heavy metals	[..........]
	(f) Dissolved silica from water attack of concrete	[..........]
	(g) Herbicide and pesticide residue	[..........]
	(h) Unusual chemicals derived from people's medication	[..........]
163.	Impervious areas return 100% of rainfall impinging on them to the urban drainage system.	[..........]
164.	Urban areas commonly produce two streams of water. These are recycled wastewater (the tailwater of sewerage plants) and stormwater. Grade the following statements as true or false.	
	(a) Wastewater recycling has the advantage of a steady supply.	[..........]
	(b) Stormwater is typically much worse quality than wastewater and is thus harder to treat.	[..........]
	(c) Storing stormwater is easier because you can predict flow volumes far into the future.	[..........]
	(d) Wastewater is difficult to treat because of its relatively high salinity.	[..........]

Answers

1. **False**. Metres, kilograms, seconds are the Système Internationale Units.

2. **True**. It is the basis of some HEC models. HEC was originally an abbreviation of the 'Hydrologic Engineering Center of the US Army Corp of Engineers' but the acronym HEC is usually used.

3. **False**, although we usually have no other option than to assume that is the case.

4. **True**.

5. **True**. Turbidity is useful because it is correlated with people's perception of water quality.

6. **False**. Turbidity is an intensive quantity; hence multiplication by a volume gives a meaningless result.

7. **True**. Very pure water gives a reading of around 0.2 NTU.

8. **True**. In many catchments the water is of similar purity to lower grades of laboratory-purified water. This means that many instruments cannot obtain a meaningful result because the water is below the detection threshold. This data is sometimes referred to as 'censored'.

9. **False**. The soils are commonly anisotropic and have many levels of variation, ranging from stochastic to systematic variation. It is difficult to measure such properties and hence we make such simplifying assumptions.

10. **True**. The noise emitted by water turbulence has a broad spectrum of frequencies with no particular frequency dominant. Hence there is no 'pitch' and it is close to a 'white noise'.

11. **True**. Even going over the highest waterfall will only heat water a fraction of a degree.

12. **False**. The water level decreases because the total energy remains constant but the velocity increases, leading to a decrease in the pressure head and hence a fall in the water surface.

13. **True**. Energy has dimensions of $M L^2 T^{-1}$ and weight has the dimension of $M L T^{-1}$. Thus, energy per unit weight has the dimension of $[L]$. This often assists visualisation of the hydraulics.

14. **False**. As the soil dries the hydraulic conductivity dramatically decreases because the water is forced to pass through smaller and smaller pores.

15. **True**. Forest soils often are better modelled as two-part media (pores and large voids).

16. **True**. Many other factors will also influence the result, but a rising hydrograph has a very different 'sediment rating curve' to a falling hydrograph.

17. **False**. Interception ranges from from 10% to 30% in a range of Australian studies but has been shown to be up to 65% in European alpine studies.

18. **True**. They are a form of graphical hydrologic modelling.

19. **True**. This can be very onerous to obtain in high forest and/or steep country.

20. **False**. It is usually around 2-3% for a 1% change in annual rainfall. A consequence is that a decrease of annual rainfall causes a much greater fractional decrease in annual streamflow.

21. **True**. The authors are not aware of any other natural cyclic variations in common circumstances.

22. **True**, although high flows may dislodge these and allow sediment movement.

23. **True**. The hydraulic conductivity is the parameter.

24. **False**. This view was held until about the 1960s, after which it was deposed by many field studies.

25. **True**. Karst watersheds are notorious for underground connections via caves and channels.

26. **False**. Rainfall intensity over a short period is only one factor of many. The total amount of rainfall is the best indicator.

27. **False**. Rainfall intensity as a function of time is a hyetograph.

28. **False**. The observations of Hurst in Egypt and since then, many other scientists have shown there is a small correlation between rainfall in successive years. This has major consequences for the behaviour of long-term records.

29. **False**. Variations in viscosity with temperature will have a major impact on accuracy, but data is rarely corrected for this.

30. **False**. One can use daily, weekly, or monthly data. Long calibration periods do not usually add more information than shorter periods and may be incompatible with human and organisation requirements.

31. **False**. The distribution is usually highly skewed towards the high flow end.

32. **False**. The water is often of a high physical quality but has high bacterial levels associated with faecal contamination by fauna.

33. **True**. It is difficult to get random samples, the underlying distributions are not normal, and there are a range of systematic variations that should be taken into account. This appears to worry some more than others.

34. **False**. The sediment concentration usually reaches a maximum very early in the event and then diminishes as available sediment is moved from the stream bed.

35. **False**. They are an imbricated (i.e. overlapping like roof tiles) and systematic natural structure and resistant to normal levels of flow erosion. There is little 'randomness' about them.

36. **False**. Streams deposit sediment on the inside of bends. This tends to make the bends more pronounced. Over time a straight channel will become a meandering channel.

37. **False**. Woody debris in streams is an important component of natural stream structures.

38. **True**. Most liquids contract when they freeze. If water did this then the ice would sink and the entire water body would ultimately freeze, to the detriment of fauna. Because the water expands the ice floats and acts as an insulator, allowing fauna to exist under it.

39. **False**. Water is very erosive because of its momentum and its ability to dissolve matter, and natural streams are constantly eroding the landscape. In most cases humans are powerless to stop such processes.

40. **False**. His observations were astute, but were not quantitative, there were no hypotheses tested, he did not relate observations to other physical phenomena, and he did not experiment. He is regarded as a 'polymath' rather than a scientist.

41. **False**. It is the other way around – the transfer of water molecules from the solid state to the vapour state without passing through being water. It is an important process in removal of ice from snow and glaciers.

42. **False**. Snow is a porous material and as the snow ages the water content increases.

43. **False**. Turbulence around the perimeter usually results in greater accumulations of snow in the openings. As the snow melts these accumulations become more visible.

44. **False**. Laminar water flow hardly exists other than in laboratory situations.

45. **False**. About 10% agreement is usually viewed as satisfactory. This reflects that rainfall is highly variable in space and time. Turbulence at the gauge entry will create large differences.

46. **False**. Many trees have 'sinker' roots which penetrate into fissures in 'bedrock' and return water to the tree in dry periods. These may be many metres below what would be classed as 'soil'.

47. **True**. The words 'effluent' and 'influent' are sometimes used for this behaviour but are better avoided since many scientists (including the authors) get them confused.

48. **True**, although the word is better avoided because it has no strict definition.

49. **True**. The driving force is that of gravity.

50. **True**, and using small bore holes the phreatic surface can be approximated by the water levels in the holes.

51. **False**.The deeper piezometer indicates that the water there is at higher energy than the shallow piezometer. Hence the water would be moving upwards, with some energy being lost to friction.

52. **False**. The action of pumping lowers groundwater around the bore dramatically. Thus, it is not a good monitor of local groundwater levels.

53. **False**. When an nth-order stream meets an nth-order stream, the order of the new stream is $n + 1$. Thus, the new stream here is fifth-order.

54. **True**. 120 mm is 1.2 ML ha^{-1}. Then volume of rainfall is 46 x 1.2 ML = 55.2 ML. Since 50% runs off then volume is 27.6 ML.

55. **False**. Net velocity is (2500/3600) m s^{-1}. Then flow rate in m^3 hour^{-1} is 2.5 x (2500/3600) x 3600 = 6250 m^3 hour. It's always those pesky zeros that muck you up.

56. **False**. The hydraulic gradient is 2/1000 = 0.002 = 0.2%.

57. **True**. Under the principle of continuity, the product of the velocity and area of cross-section are constant.

58. **False**. It usually indicates an obstacle such as a rock. The water accelerates (cross-sectional area decreases so velocity increases). This means that energy is transferred from height to kinetic energy (pressure at the surface is constant) and so the surface dips.

59. **True**. This is often used as a local indicator of rain the next day.

60. **True**. The geometry of rainbows demands this for the first and second harmonic rainbow (the only two commonly seen).

61. **True**.

62. **False**. Continuity means that the volumetric (or mass) rate of product is maintained. Thus for (incompressible) water, the product of velocity times area of cross-section is constant.

63. **True**. Often used to attempt to limit headward erosion in stream gullies.

64. **False**. The water may be 'cleaned up' but it often comes at the price of enhanced downstream erosion.

65. **True**. Generally, it implies a wet field.

66. **False**. Most wetlands have varying water levels and many wetlands need a 'dry' period for species to complete their life cycle.

67. **True**. Effective upstream works combatting erosion can often lead to degradation of downstream wetlands by entrenchment.

68. **False**. The Froude Number is dimensionless. It is a measure of whether downstream effects on flow can be transmitted upstream.

69. **False**. The pressure gradients can 'pull' water but, in unsaturated soil they can't 'push' water – somewhat analogous to pushing a piece of string.

70. **False**. Ideally the model parameters should be measured independently. Calibration by optimisation on a few parameters may achieve a good model fit, but the physical reality of the model is discarded.

71. **False**. *Giardia* and *Cryptosporidium* are not removed by chlorination and this can be a water supply threat.

72. **True**. The wolves are reducing elk-browsing, allowing aspen and willows to regrow. This favours beavers. The work is viewed as 'classic hydro-ecology'.

73. **False**. Riparian and flood plain forests thrive on some flooding. River regulation reduces this vital input of water and nutrients.

74. **False**. Many studies have shown decline in these forests associated with river regulation, water diversions, and dam construction. This indicates that they are a flood-dependent species.

75. **False**. If they do not get regular flooding there is usually a long-term decline but in human terms this is slow.

76. **True**. It is difficult to separate the two effects.

77. **False**. Large floods (and no floods) are usually fairly constant as to where the water is (i.e. everything is under water). The middle range of flooding gives much more variable results as to which areas flood and which don't.

78. **False**. Flooding forests would usually be viewed as a subset of riparian forests.

79. **True**. These include specialised tissue to take in and translocate oxygen to the roots. Floating roots are a common adaptation. Swamp cypress includes 'knees' that protrude above the soil (their full function unknown).

80. **False**. Often the water level may be such that areas that once only flooded intermittently are permanently inundated. It usually changes the variation in flooding.

81. **True**. Historically associated with oil exploration.

82. **True**. It thrives on flooding but is very drought tolerant and occurs over much of Australia. This was noted by early botanical explorers and it led to the tree being planted all over the world in 'difficult' sites.

83. **True**. The terms often have distinct local meaning.

84. **False**. All the processes contribute to water excess, and it is impossible to separate such processes.

85. **True**, and 'lacunarity' is sometimes used to describe a landscape with lots of hollows.

86. **True**. Pressure $= \rho\,g\,h$ where ρ = density (1000 kg m^{-3}), g = 9.8 m sec^{-2}, and h = 2 m. Thus pressure = 19.6 kPa.

87. **False**. It is probably unmeasurable for all but the most well-defined channels. In headwater streams the perimeter is essentially fractal in nature.

88. **True**. The ability to use these equations in field hydrology improves with practice and experience. A difficulty is that the parameters differ with the units used.

89. **False**. Scientists from different disciplines will disagree on the answer, but it is usually viewed as many tens of metres below the stream and perhaps hundreds of metres below the stream.

90. **False**. The hydraulic conductivity of bedrock is usually dominated by fractures rather than the conductivity through the matrix. Small samples do not include this feature. The question of the 'bottom of watersheds' (and whether they actually have one) is an active area of research.

91. **True**. But that covers a wide variety of rock types and permeabilities.

92. **False**. There are a number of such equations but they tend to only be reliable within restricted sub-categories.

93. **False**. Under the 'Ramsar Wetland' convention the wetland must be managed to conserve wetland values by maintaining the status quo of past management.

94. **False**. A cut-off meander is formed naturally when the main course of a river moves, leaving its old course remote and generally unconnected to the new course.

95. **True**. There are many other local names.

96. **True**. These have been commonly used in the last century.

97. **False**. Levees seem to carry with them a horrendous collection of economic, political, social, and ecological baggage.

98. **False**. Other way around – 1.5 units horizontally to 1 unit vertically. Steeper walls fail prematurely.

99. **False**. It has been shown that the banks deflect under heavy vehicles and this leads to bank failure during floods. Many organisations place bans on vehicle traffic along the banks for this reason.

100. **True**, but for all the wrong reasons. Many failed, and some were 'dynamited' to allow flooding of lower-value property and save higher-value property from flooding. The failures highlighted strong and unresolved social inequity.

101. **True**. These can often recolonise the main channel at times of hydraulic reconnection.

102. **False**. Wetlands near the coast may suffer influences of tides including higher water levels in discharging streams and variations in groundwater pressures.

103. **False**. Usually the wetlands have formed because of upstream erosion leading to alluvial sediment. Reduction of upstream erosion can destabilise such wetlands.

104. **False**. Floods are just part of the hydrologic variation. Because of poor and often historic land-use decisions they may well lead to human, economic, or social disasters but in themselves they are not hydrologic disasters.

105. **False**. It is the opposite and refers to oxidation of organic compounds when the soil is unflooded. The loss of these causes the land surface to fall (relative to sea levels) appreciably.

106. **True**. Settlement on such soils has issues of increased flooding and property damage associated with infrastructure settlement because the land is lower.

107. **True**. The levee pushes the flood waters to the other side, and also reduces the area of flood plain available so increases flood levels. Historically this led to outbreaks of 'levee dynamiting'.

108. **False**. You may get some elevation of land surfaces, but there is no guarantee that the areas that deflated will be the areas that elevate.

109. **True**. (more or less). There was a wave of land clearing in mountain regions. This led to much philosophical speculation and some early ventures in hydrology measurement.

110. **True**. The alpine areas made famous in pictures such as *The Sound of Music* are not proving attractive places to live for the modern generation. This appears to be associated with a reduction in streamflow as the areas become forest.

111. **True**. The war led to the abandonment of cotton cultivation and many of the areas reverted to forests. These transpired more water than the agricultural land, giving diminished stream flows.

112. **True**. The water yield is high for young trees, then diminishes to a minimum at about age 40, and then starts increasing again. This is known as the 'Kuczera effect' after an early attempt to quantify it.

113. **True**, although (fortunately) rare. The Vajont Dam disaster (Italy, 1963) was an example of this. This has been described as an 'inland tsunami'.

114. **False**. All rock is permeable to some extent due to fractures. In particular, upper layers of bedrock are often oxidised and highly fractured. It has been shown that the rock zones do transmit 'stormflow' but more research in this area is needed.

115. **True**. The relation between water level ('stage') and flow rate is referred to as a rating curve.

116. **False**. The measurements are usually made in a 'well' to reduce the effect of surging from water turbulence. In electronic data logging this can give erroneous readings. In the older paper-and-pen recordings, the surging would give thick, wet pen traces that would run the pen out of ink.

117. **False**. A data logger records data in some readable form. The paper chart recorder meets this criterion.

118. **False**. If the study goes on for many decades, the 'control forest' will also change. It is arguable whether there can be a 'true control' in such situations, or what 'a control' means.

119. **True**. The one caveat is that the oxygen and hydrogen are often created by the electrolytic decomposition of water, so it is unlikely that there is any net gain in the amount of water in the world.

120. **True**. However, a word for water appears in virtually all languages.

121. **True**. There is a theory that much of it came from comets bombarding the earth, but this is far from 'proven' or generally accepted.

122. **False**. The chemical 'water' may occur as ice on some planets and traces of water vapour have been found in space. It is thought that liquid water may exist occasionally underground on some planets, but this is not proven.

123. **True**. Volcanic emissions often contain substantial amounts of steam, and ejected ash material may include water trapped inside

solidified 'ash'. But there is no evidence of it having anything analogous to a 'tank of water'.

124. **True**. These give a very distinctive dendritic network that is more or less fractal in nature.

125. **False**. Erosion is a natural process, although it may be inimical to human property values. It is arguable as to whether it can really be stopped.

126. **False**. It is the other way around. Both are a result of the surface tension of water and are unusual properties of the liquid 'water'.

127. **False**. Water is more polar than most comparable liquids and it is thought that this derives from hydrogen bonding. This makes it much more 'adhesive' and cohesive, and leads to a vast reactivity. It is unusual for liquids to expand when they freeze (as water does), and water has an unusually high heat capacity.

128. **True**. It is not known as 'The Universal Solvent' for nothing. Thus erosion is part mechanical action, part dissolution.

129. **True**. Some have argued that the 'virtual water' or goods should be a consideration in trade.

130. **True**. The value is usually derived from some sort of 'modelling' rather than direct measurement.

131. **False**. Firstly, that piece of meat has already been produced. Secondly, in the long term some other land use would replace the meat-growing and that may have a similar water use profile. There can often be considerable passion concerning this question.

132. **True**. However, they have different uses, are expressed in different ways and units, and are derived in different ways. The photosynthetic efficiency usually refers to very small amounts of matter. The virtual water content refers to much larger volumes of commercial products. There is (or should be) a 1:1 correspondence if such factors are taken into account.

133. **True**. Many areas within the desert have no measurable rainfall, although clouds may form. Some areas are quite well supplied with groundwater by the Agua Verde Aquifer and can support limited human endeavour. The area has been used to simulate 'being on Mars' in scientific and film endeavours.

134. **False**. It needs at least to be quantitative. There are a number of drought indices or techniques that can be applied.

135. **True**. In such situations, definitions such as 'rainfall below average' become meaningless.

136. **False**. Many economic studies show that although irrigation is politically popular, the economic gains from such 'headworks' usually doesn't pay the cost of construction. Although this is well known, countries still expand their irrigation capacity. Presentation of such a result is guaranteed to start a debate in irrigation areas.

137. **False**. Many centre pivot systems are fitted with supplementary irrigators to give a square pattern. This reflects that paddocks are rarely circular.

138. **True**. Irrigation without drainage is a recipe for disaster, and the drainage developments came relatively late in the history of humanity.

139. **False**. This was an early strategy but is now usually illegal. Amongst other things the water may carry a high pesticide loading.

140. **True**. The incident led to the definition of the 'Kesterson Effect' after the name of the wetland. Amongst other things the waste carried high concentrations of selenium which were a cumulative toxin to birds. It led to a worldwide rethink of what to do with irrigation drainage.

141. **True**. Probably from the Latin *'potare'*, to drink.

142. **False**. Crystal-clear water is a good start but pathogen counts attributable to fauna (particularly *Giardia* and *Cryptosporidium*) can be high.

143. **False**. Pigs, followed by cattle, followed by sheep and goats. This can lead to prohibitions against pigs in municipal watersheds.

144. **True**. This reflects that many organisms that affect pigs also affect humans and vice versa.

145. **False**. The three pipes carry potable water, low quality water for gardens, and sewerage. These systems have proven to be expensive and difficult to manage.

146. **True**. Most stream ratings and streamflow data is collected by hydrographers.

147. **True**. Sometimes to achieve this desirable state, considerable 'smoothing' work is needed at the measurement station.

148. **False**. Unless the stream is passing through a 'critical flow' (i.e. a drop) at the rating cross-section, what happens downstream will have a big influence on the rating. Hydrographers try to minimise this by finding the most stable sections of channel available to use in their rating. Often gaugings are viewed as a necessary compromise.

149. **True**, although the former implies that moisture levels are quite high compared to the dryer parts of the unsaturated zone.

150. **True**. Studies have shown that they often exist as a transient state after rainfall and may reflect macropore transmission into a non-macropore zone. Additionally, the term can be used when there are multiple but separate layered aquifers separated by aquicludes.

151. **True**. The old root mass decays and the new root mass has not developed adequate anchoring support to sustain the hill slope.

152. **True**. The root mass may sustain the bonding of surface soil to the rock but the weight of the growing trees may overload a weak slope. Failure may occur well below the root zone.

153. **False**. Flowing traffic and water is modelled quite well by 'kinematic wave theory' but there is nothing analogous to traffic lights in streams.

154. **False**. Sometimes (particularly in karst landscapes) one might be able to make some accurate surmises, but usually (and particularly in weathered landscapes) there is no clear link between the type of underlying bedrock and the behaviour of streams.

155. **True**. The difficulty is that we have neither the measurement technology nor the computational methods to deal with such issues.

156. **True**.

157. **False**. If the evaporative demand exceeds the precipitation the system is water limited.

158. **False**. Because of the relatively deep, porous soils, water from previous events is stored and 'pushed' out by the 'new' water through the soil. Overland flow is unusual in forests.

159. **False**. Erosivity is a rainfall property. Erodibility is a soil property.

160. **True**.

161. **True**.

162. (a) **True**. Lots of mechanical disturbances

(b) **False**. Little mercury waste these days.

(c) **False**. Silver is less used in photography in the digital age, and most silver is recovered in commercial plants. Residues would end up in sewerage plants, not as urban runoff.

(d) **True**. Fertilisers, faecal matter, etc.

(e) **True**. Particularly from corrosion of metal structures.

(f) **False**. Silica is found in natural waters but as a low-level natural occurrence.

(g) **True**. Reflects widespread garden use.

(h) **False**. Such residues would be found in sewage.

163. **False**. Usually about 90% if returned – the 10% is known as 'initial loss' and reflects depression storage (puddles), evaporation, and wetting of surfaces.

164. (a) **True**. This allows elaborate plants to be built.

(b) **False**. Pollutants are at a much lower level, less offensive, and many settle out.

(c) **False**. Stormwater is highly variable in space and time.

(d) **True**. A major source of treatment is dilution with fresh water in a river or stream.

5 Image-based Questions

1. Heavy Rains

The plot below is courtesy of the Bureau of Meteorology Australia.

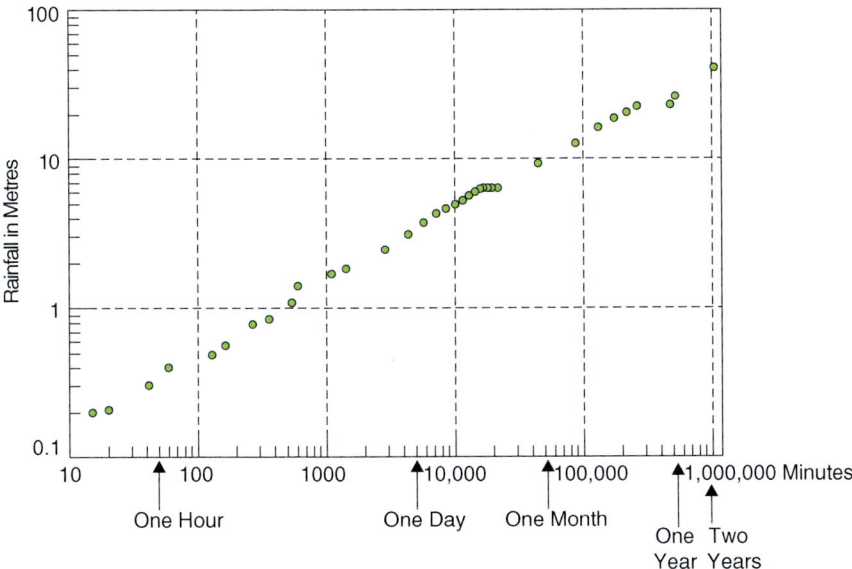

Fig. 5.1. Held under 'Creative Commons', supplied by the Bureau of Meteorology, Australia.

A. Look at the two axes and note that the scale contracts away from the origin. What is this type of scale called? Why is it used?

B. For 1 hour the maximum rainfall intensity is about 0.4 m (400 mm). What would the weight of water (in kg) be falling on 1 hectare in this time?

© Leon Bren and Patrick Lane 2021. *Key Questions in Hydrology and Watershed Management* (L. Bren and P. Lane)
DOI: 10.1079/9781789249682.0005

C. The top 10 points are from Cherrapunji (aka Sohra) in India and recorded in the 1860s. Can you find out anything about this place?

2. Raster Hydrograph

In this the entire flow record of a long period at short intervals can be represented by colours (see example below). In this example:

A. There is an arbitrary choice of colours with red representing low flows and blue representing high flows. Is this a good choice? What makes a poor choice?

B. The scale appears to be logarithmic in nature. What do we mean by this? What are the advantages and disadvantages of this?

C. Suppose you had composed and printed this but wanted to make it a 'truer' representation. What might you do that would be more geometrically correct? Hint: the last day of December passes into the first day of the following year.

D. What changes suddenly appear about 1961. Without knowing anything about the specifics of the gauging station, suggest three things that might account for the change (and that could be further investigated).

E. What alternative forms of display could be used for data in this form?

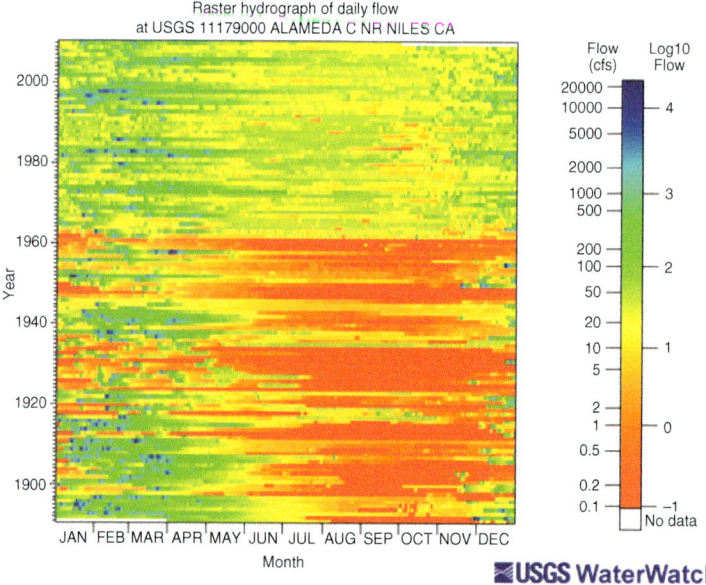

Fig. 5.2. Courtesy of the US Geological Survey.

3. Boundary of the Watershed

A. On the following map, mark out the boundary of the watershed contributing to point A.

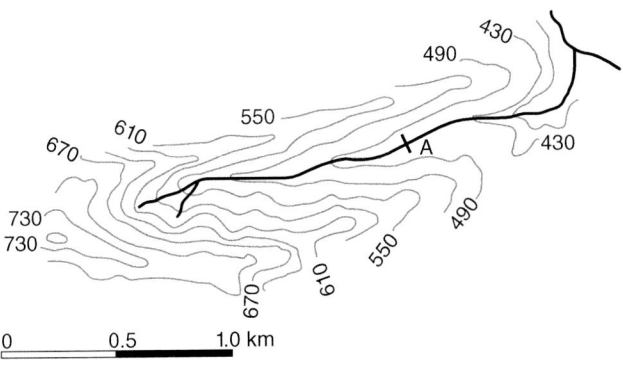

Fig. 5.3A. Reproduced courtesy of Springer Nature.

B. On the map in Fig. 5.3B, mark out the boundaries of the watersheds above Points A, B, C, and D. The map-maker has had a problem near points A and B – can you see what it is?

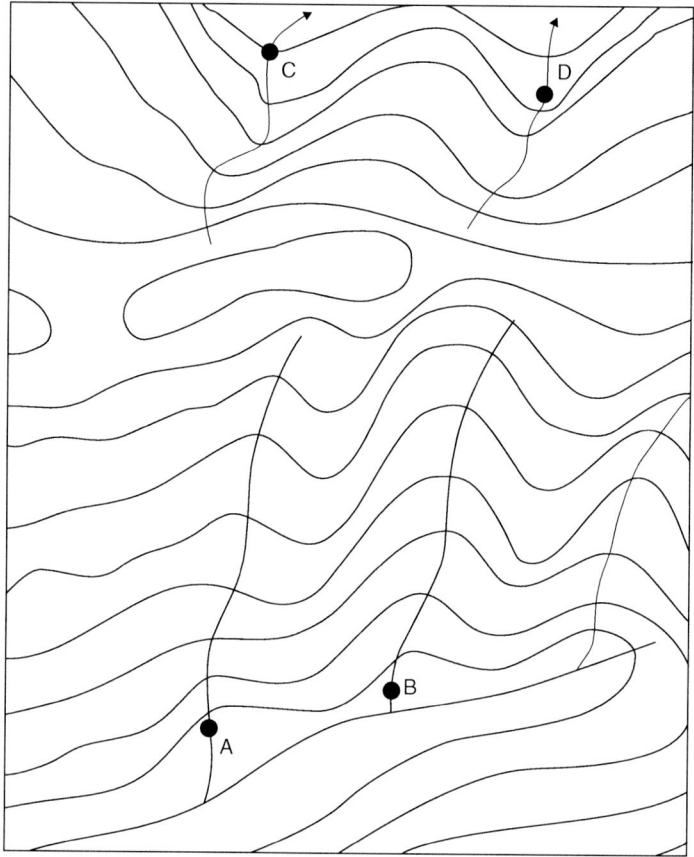

Fig. 5.3B.

4. Watershed Boundaries on a Real Contour Map

A portion of the 1:50000 Mansfield (Victoria, Australia) map is shown. The map scale has changed but the grid is 1 km x 1 km and the vertical orientation is magnetic north. Contour interval is 20 m. Delineate the watershed boundaries at points A, B, and C. Note that the boundaries may go outside the portion of map given. Hint: do the left-hand and right-hand boundaries separately and see where they join. Note: it is suggested that students find contour maps of their home area and practise delineation – it's a useful skill that dramatically improves with practice.

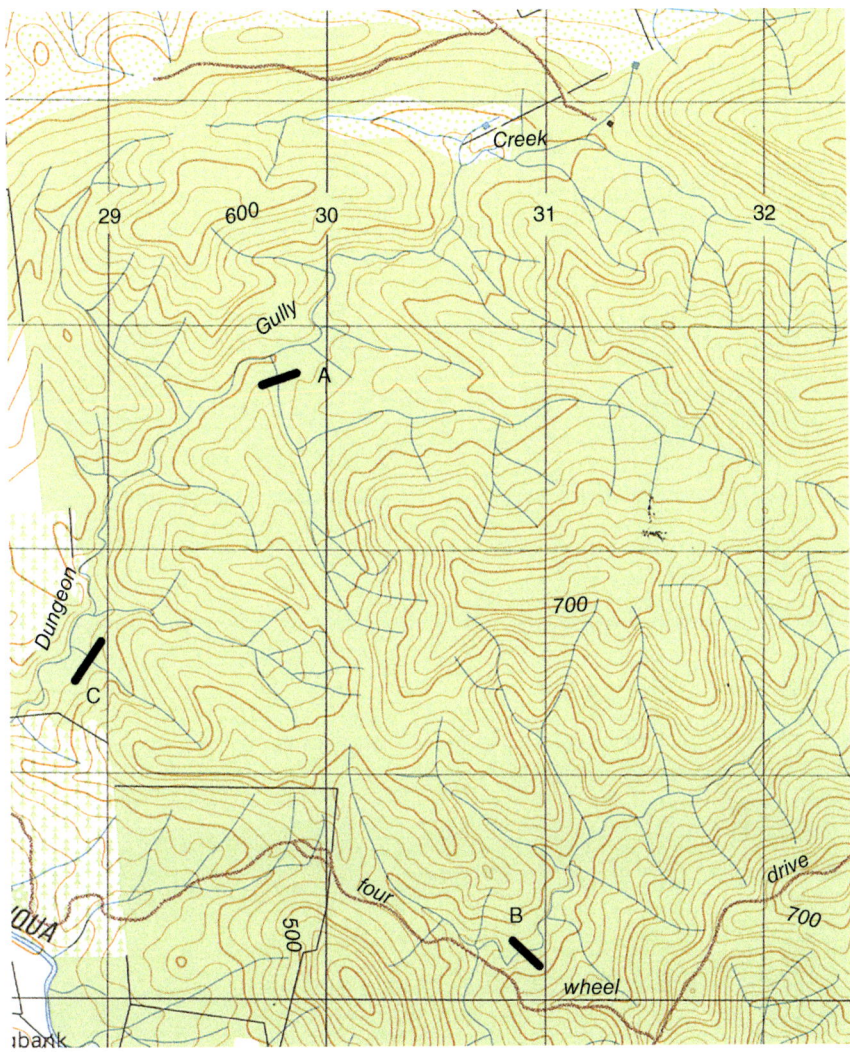

Fig. 5.4. Map courtesy of VicMaps and the Department of Land, Environment, and Water, Victoria, Australia.

5. Strahler Ordering (1)

A. On the map in Fig. 5.5.A name the Strahler stream orders. Hint: the smallest streams are Order 1. The stream flows from the top of the page towards the bottom.

137

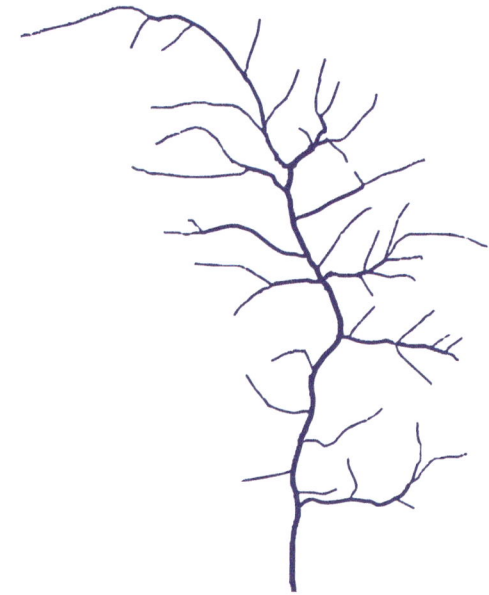

Fig. 5.5A. Reproduced by courtesy of Springer Nature.

B. Strahler Ordering (2)

The rules of Strahler ordering are usually written as
Rule 1: n + <n -> n
Rule 2: n + n -> n+1
In the diagrams in Fig. 5.5B, give the stream-order and note the rule applied.
Note that one case has some ambiguity.

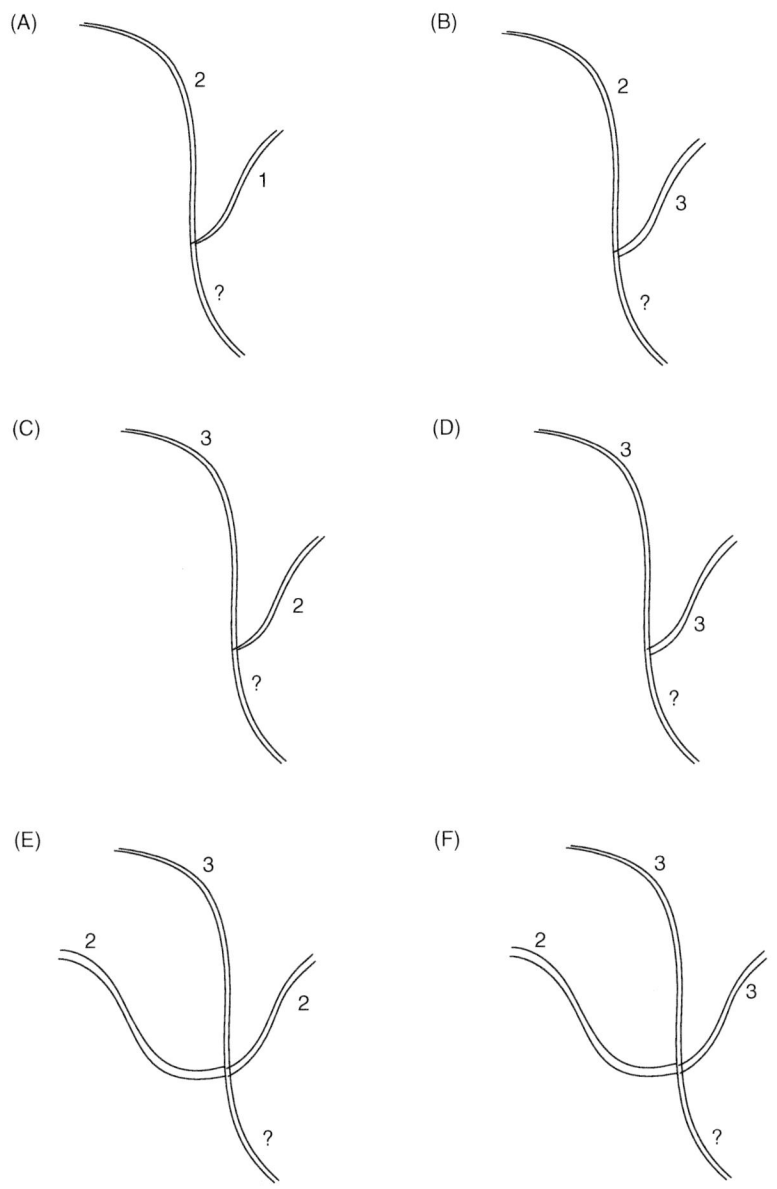

Fig. 5.5B.

6. Big Floods and Bigger Floods

It is hard to get a handle on the hydrologic magnitude of floods worldwide. One paper that has done this is O'Connor et al. (2002). The two illustrations are based on this. Study these and answer the following questions.

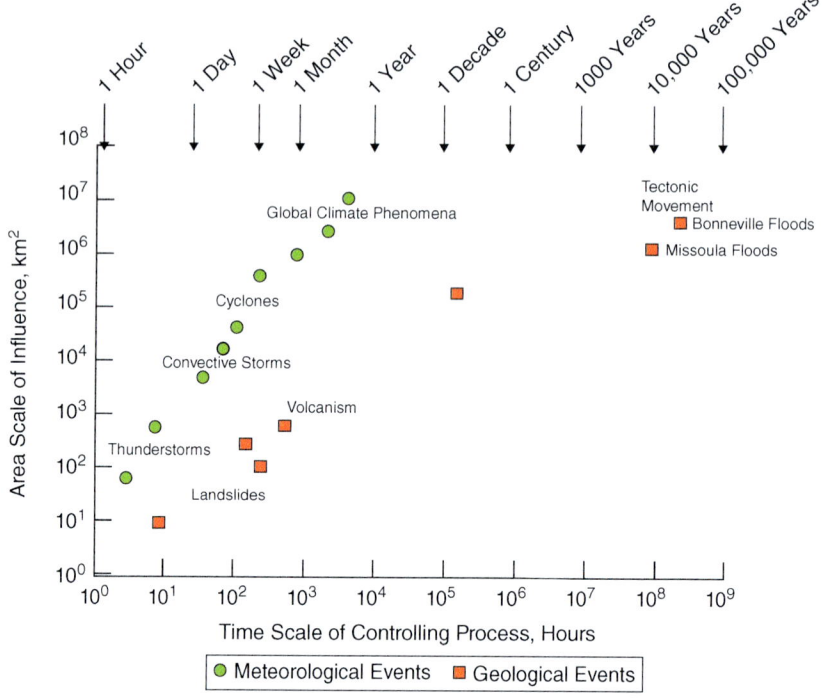

Fig. 5.6A. Space-time domain for selected meteorologic and geologic processes that can cause floods. Selected floods are also shown (some named), and the type of event are also indicated.

From Fig. 5.6A.

A. What do we mean by floods from 'Geological Events' as distinct from floods from 'Meteorological Events'?

B. The Missoula (Montana) floods appear to have lasted for 15,000 years. Have we made a mistake here?

C. The meteorologic event points come largely from the USA. Does this mean that the USA is an unusually flood-prone country?

D. Can you think of a local flood and place it on the plot? What factors lead to floods being particularly destructive in economic terms?

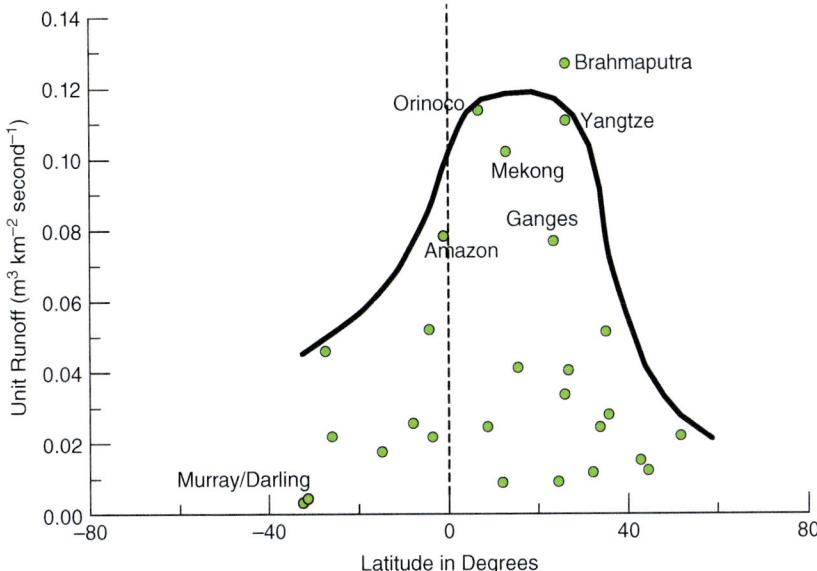

Fig. 5.6B. Largest recorded floods from basins larger than 500,000 km² plotted as unit discharge (peak discharge per square kilometer per second), and the 'limiting line' postulated by O'Connor *et al.* (2002). The points shown are either rainfall events or rain/snow events.

From Fig. 5.6B.

This shows the flood data and postulates an upper limit for rain and rain/snow events. The data on which it is derived is also shown with some of the points 'named'.

E. The limit line is asymmetric about the equator and, in the southern hemisphere, does not extend to the higher latitudes shown for the northern hemisphere. Why is this?

F. Compare the floods marked for the Murray/Darling Rivers in Australia and the Yangtze River in China. The former are sparsely populated watersheds compared to the latter. The flooding in the lower latitudes (i.e. nearer the equator) can be very large. What might one say about these areas that are subject to such large floods?

G. Fig. 5.6B. has excluded 'snowmelt-caused flooding' in the northern hemisphere which lies outside the limiting line. Why do we not get such snowmelt-caused flooding in the southern hemisphere? Do we get the same amount of snow in the southern hemisphere as in the northern hemisphere but it just falls into the ocean and melts?

Figs 5.6A and B redrawn from original papers courtesy of the US Geological Survey and the US Forest Service. The contribution of the authors is noted.

7. Impacts of Burning on Watersheds

The illustration below was presented at an Australian Forest Hydrology Conference and shows the instantaneous hydrograph when a small research watershed was burnt by an intense forest fire in summer. Match the four labels with the features on the hydrograph.

1. Time of burning

2. Spike flow due to burning (one of three)

3. Diurnal variation

4. Increase in flow after burning due to cessation of transpiration

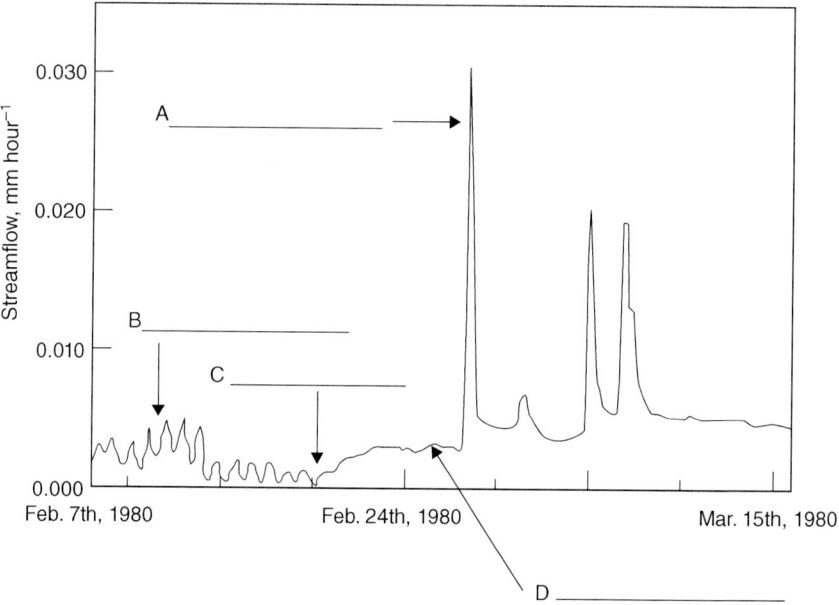

Fig. 5.7. Reproduced courtesy of Engineers Australia.

8. Properties of Water

What properties of water are being shown by the illustrations? Also answer the supplementary questions.

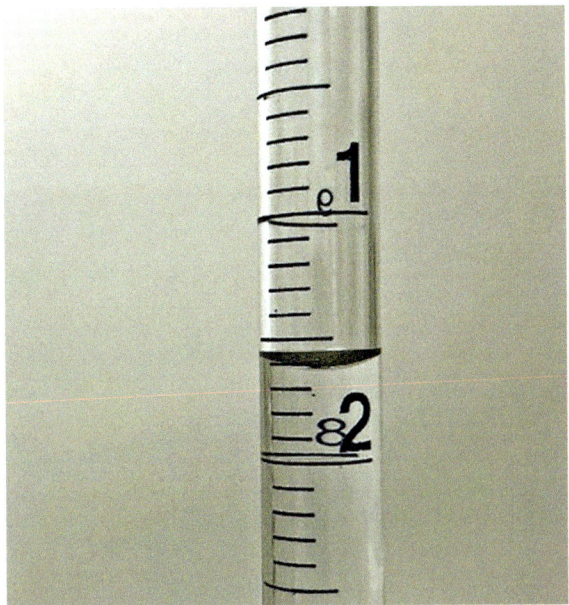

Fig. 5.8A. Where does one read the pipette?

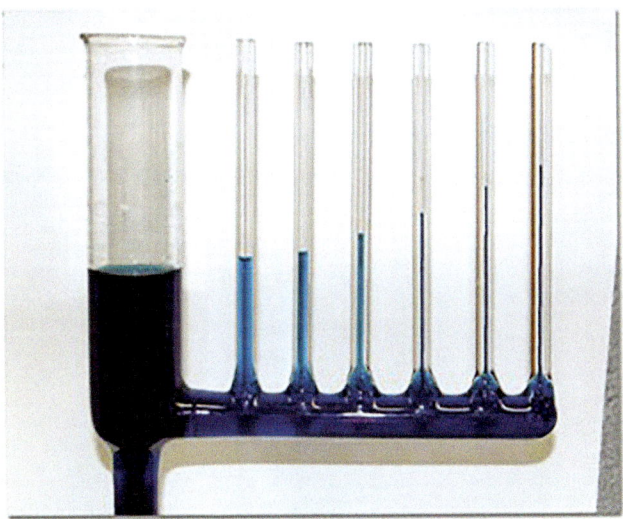

Fig 5.8B. What concepts here are sometimes found in soil physics?

Figs 5.8A and B. Courtesy of the US Geological Survey.

9. Water to Ice

We have stressed that water is unusual in that, due to hydrogen bonding, as it freezes it expands, and thereby achieves a lower density. Below is a picture of an ice cube that sinks in ordinary water, not floats. How can this be?

Fig. 5.9. Courtesy of the US Geological Survey.

10. A Common Water Supply Issue

The glass and the tea spoon have been washed but the result is not all that is desired. What is the problem?

Fig. 5.10. Courtesy of the US Geological Survey.

11. Hydrometric Devices

The images in Figs 5.11A–H show standard measuring devices sometimes encountered in hydrology. Name them and give a brief description of their role in hydrology science.

Fig. 5.11A.

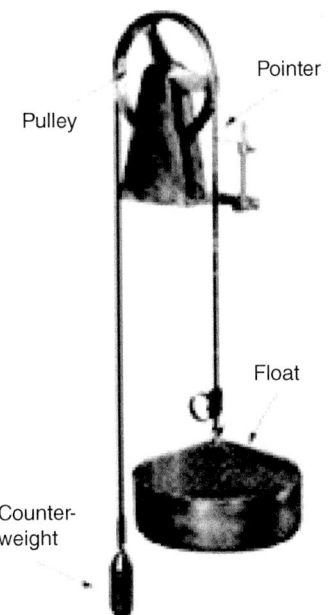

Pointer

Pulley

Float

Counter-weight

Fig. 5.11B.

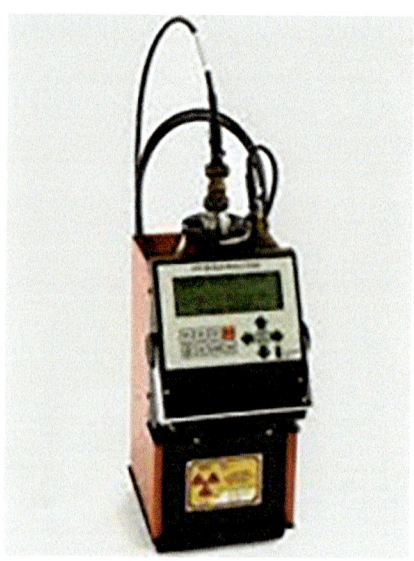

Fig. 5.11C.

Fig. 5.11D.

Fig. 5.11E.

Fig. 5.11F.

Fig. 5.11G.

Fig. 5.11H.

Figs 5.11A, B, D, E, G, and H. Courtesy of the US Geological Survey.
Fig. 5.11F. Courtesy of Campbell Scientific Australia.

12. How Things Used to Be Done

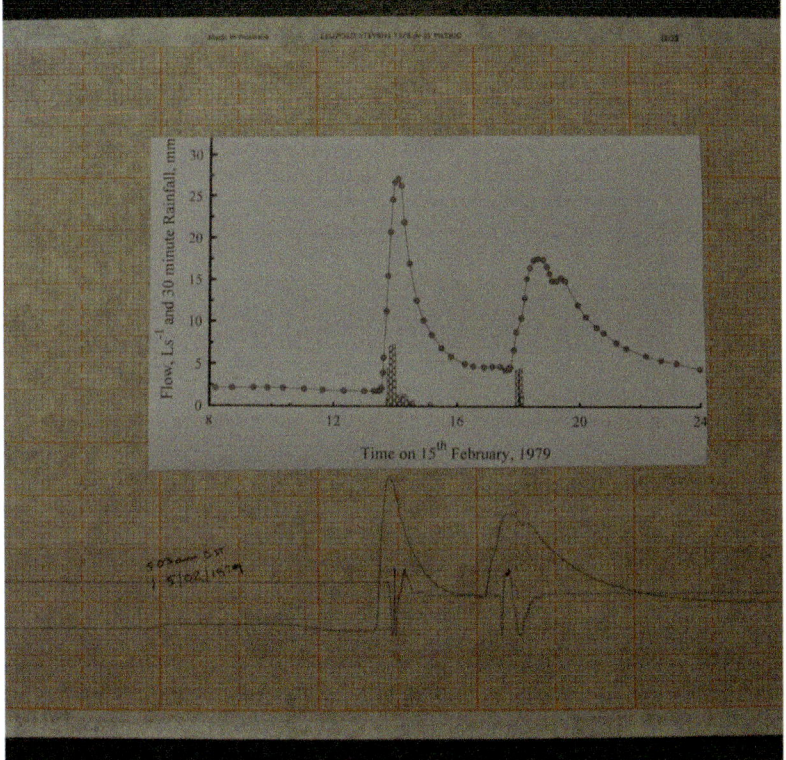

Fig. 5.12.

The upper image of Fig. 5.12 shows 45-year-old charts awaiting yet another analysis. The lower image shows a chart record of a small storm hydrograph from the Croppers Creek Hydrologic Project in Victoria, Australia. The vertical divisions are 6 hours apart. The flow pen records the height in centimetres. A tipping bucket rain gauge makes a 'jog' on the chart when it tips. Each jog represents 0.2 mm. of rainfall. This pen has a domain of 20 tips and then reverses direction. A time mark on the chart is made by a visitor. We have digitised both records and printed them out in a more readable form. In the hydrograph the digitised points are also marked.

149

A. How would such a recorder be driven?

B. How would this chart be converted to a digital record in the modern age?

C. Is this a data logger?

D. What are the advantages and disadvantages of this form of recording compared to modern electronic devices?

13. Watershed Hydrology and Political Boundaries (1)

The sketch is adapted from an aerial photograph of the River Murray flood plain near Wodonga (Victoria). This river forms the boundary between the Australian states of Victoria and New South Wales (NSW). The greyness of the image is about proportional to the probability of water being found in the channel shown, with the river being permanent.

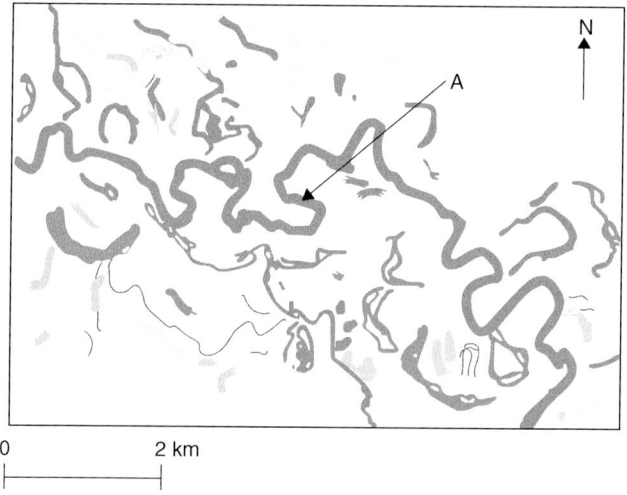

Fig. 5.13.

A. What are we looking at here?

B. Can you find 'the main course'?

C. (Horrible but true). A court decision for a murder committed on the river bank confirmed the boundary as 'the high bank of the River Murray on the southern side'. Given that definition, what is the problem at the point marked A with the arrow? Can you find other such points in the area shown?

D. A recent report has shown large 'slugs' of sand from mining works about 150 years ago are blocking the river at one of its narrowest sections. The geomorphologists concluded that the river may 'avulse' – jump to a new course that could be many kilometres from the old. Since the river is the state boundary this would effectively transfer a large area of land from one state to another and possibly put hitherto 'river' towns well away from the river. What might be done?

14. Watershed Hydrology and Political Boundaries (2): The Gambia

The Gambia is a small and narrow country whose borders mirror the meandering Gambia River in Africa. The border is based substantially on being a constant distance from the Gambia River.

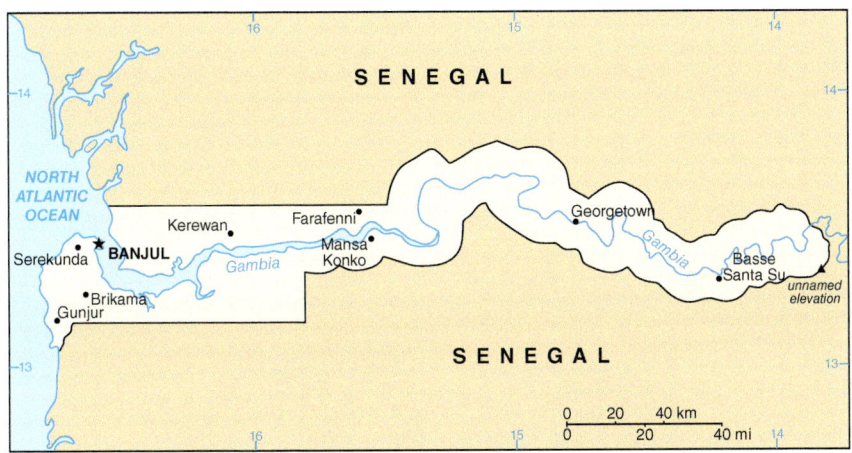

Fig. 5.14. Reproduced from Wikipedia and is held under 'Creative Commons'.

A. What are the advantages and disadvantages for The Gambia of having a country based on a riparian boundary?

B. What might the views of neighbouring Senegal be?

15. Erosional and Depositional Forms

What are we looking at in each of the photographs Fig. 5.15A–E? Answer the questions attached to each photograph.

Fig. 5.15A. Suggest an erosional and depositional sequence for this Alaskan river.

Fig. 5.15B. What is the name of this sort of deposit and how has it formed?

Fig. 5.15C. We've had a fire in the watershed, and nature is rearranging things a bit. What is going on?

Fig. 5.15D. An important and under-rated form of stream stability.

Fig. 5.15E. A landslip. What factors are at play here. Might the ocean play a role?

Figs 5.15A and **E** courtesy of Marli M. Miller (Geology Pics, Marli Miller Photography). **Fig. 5.15B** provided by Callan Bentley, Piedmont Virginia Community College.

16. What Is This?

Fig. 5.16. Held under Wikipedia 'Creative Commons'. Photo courtesy of Mark Wilson, College of Wooster.

Is this:

A. Stream channels formed across a tidal wetland as the water withdraws to the ocean at low tide (taken from a drone photograph)? Distance across the image is about 200 m.

B. Manganese dendrites on a limestone bedding plane from Germany? Distance across the image is about 20 mm.

C. LIDAR image of a stream network in an arid part of Afghanistan? The valley bottoms have been rendered black to enhance this watershed network. Distance across the image is about 25 km.

17. Landscape Evolution

Solymon and Tucker (2004) looked at the role of storm duration in the formation of landscapes. They found that long duration storms produced prolonged hydrographs that gave a much more entrenched dendritic network of streams. In contrast, short storms tended to give a much less developed landscape often evident in arid, mountainous zones. An illustration from this paper is given below.

Role of Storm Events in Shaping Hydrographs

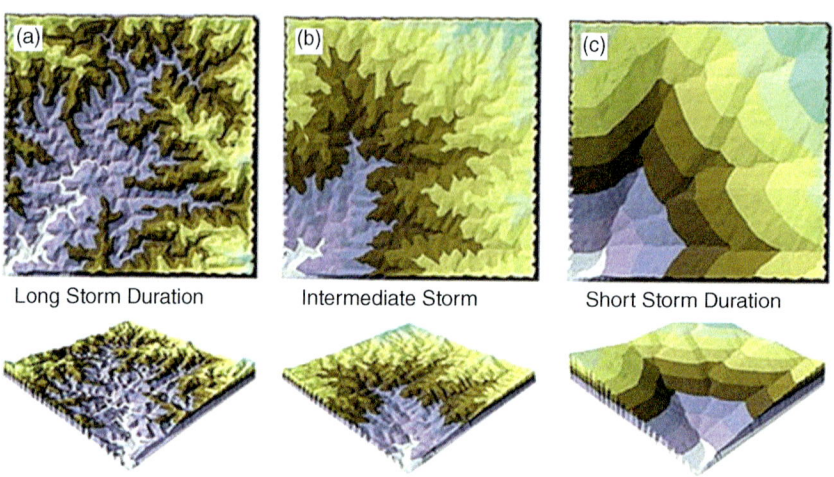

Fig. 5.17A. From Solymon and Tucker (2004): Effect of limited storm duration on landscape evolution, drainage basin geometry, and hydrograph shapes. Journal of Geophysical Research: Earth Surface, Volume 109, Issue F3.

Using this illustration, classify the four landscapes shown in Fig. 5.17B–E as being of 'long storm' or 'short storm' duration. You might extend this to landscapes with which you are familiar with.

Fig. 5.17B.

Fig. 5.17C.

Fig. 5.17D.

Fig. 5.17E.

Permission to reproduce **Figs 15.17A** and **B** provided by Gregory Tucker. The illustrations are public domain. **Fig. 15.17C** courtesy of Sharon Rasco, Sydney. **Fig. 15.17D** held under Wikipedia 'Creative Commons' courtesy of Dan Hobley. **Fig. 15.17E** courtesy of Marli M. Miller (Geology Pics, Marli Miller Photography).

18. Groundwater/Stream Relationships (Influent and Effluent Streams)

The illustrations below are schematic cross-sections of a land–groundwater–stream interface. Allocate the following labels to the illustrations below. Labels 5 and 7 are generic classifications of a class of streams. Note that some labels are common to both illustrations or are in the legend and titles, so don't forget these. The vertical scale has been greatly enlarged for illustrative purposes.

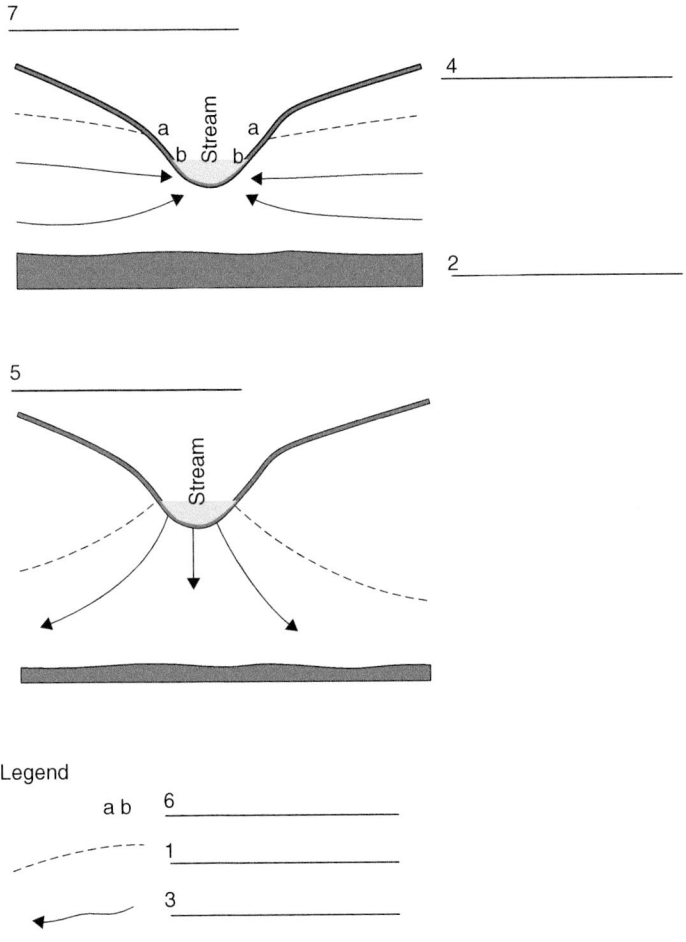

Fig. 5.18.

A. Land surface

B. Influent stream

C. Effluent stream

D. Phreatic surface

E. Groundwater vectors

F. Seepage face

G. Bedrock

19. Groundwater Pumping and Cones of Depression

The illustration is a schematic illustrating the formation of cones of depression in the phreatic surface after continued groundwater pumping. Match the labels below with the place indicators. Note that such illustrations are hopelessly out of scale. Thus, the bore may be 0.5 m in diameter and perhaps 50 m depth but the cone of depression may extend for kilometres.

A. Outflow of pumped water

B. Soil surface

C. Unsaturated zone

D. Phreatic surface (no pumping)

E. Phreatic surface after some pumping

F. Phreatic surface after extended pumping

G. Bore inlet and submerged pump

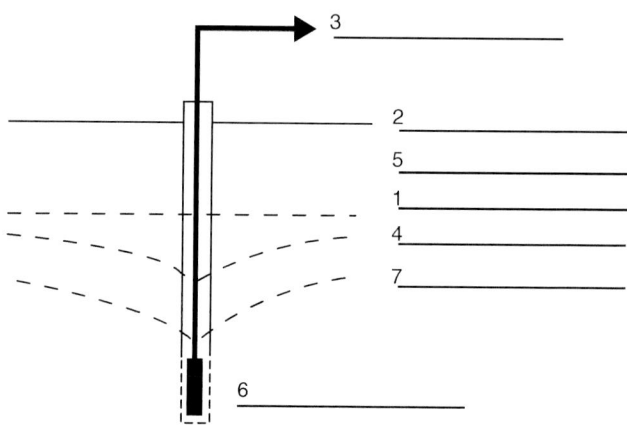

Fig. 5.19.

20. Coastal Aquifers (1)

The illustration shows a schematic of land and a freshwater phreatic aquifer intersecting the coast line. Two piezometers measure piezometric pressure. Match the lines, area, or feature with the caption:

Designation	Label or Feature
A. Line AB	(1) Land surface
B. Line AC	(2) Land–ocean–groundwater interface
C. Line AD	(3) Fresh–salt water interface
D. Line AG	(4) Salt groundwater domain
E. Line AA'	(5) Phreatic surface
F. Lens ABD	(6) Ocean
G. Area AGH	(7) Fresh groundwater domain
H. Area AGFED	(8) Ocean bed
I. Point A	(9) Mean ocean level

Would the water in the piezometers be fresh or salt?

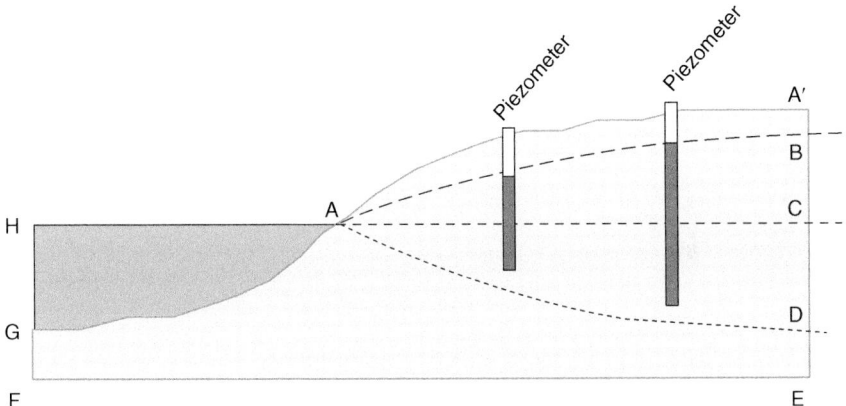

Fig. 5.20.

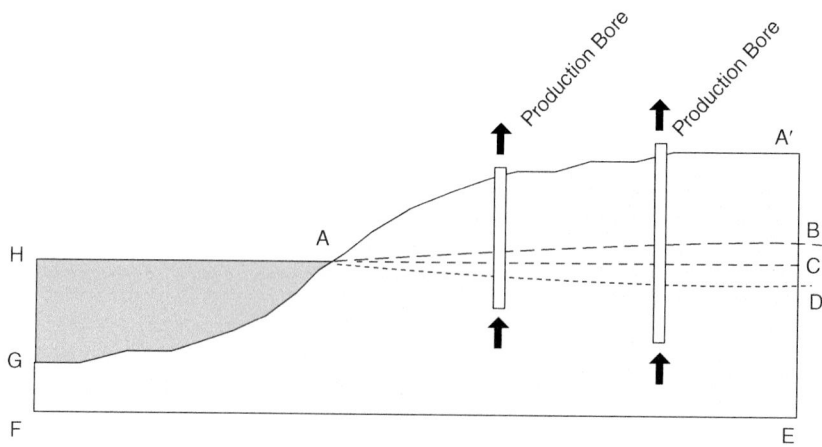

Fig. 5.21.

21. Coastal Aquifers (2)

See Fig. 5.21 above. The piezometers of the previous example have been replaced by pumping bores. The inlets and outlets of these have been marked by arrows. However, there has also been large pumping of fresh water from the aquifer inland.

A. What has happened to distances BC and CD of the previous case?

B. What important change has occurred at the inlet of the bores?

C. What strategies may be undertaken to avoid such changes in the future?

22. Interpretation of Multi-depth Piezometers

Illustrations A, B, and C in Fig. 5.22 show three multi-depth piezometers (water in black) and three possible conclusions (1 to 3). Match the data shown by the piezometers with the conclusions, giving reasons. What happens to excess energy? Does the nest of piezometers tell us about the horizontal direction of movement?

Possible Conclusions:

1. Groundwater at the depths measured by the piezometers has the same total energy per unit weight. There is no vertical component to the flow.

2. Groundwater at depth has a greater total energy per unit weight than the shallower water. Hence there is an upward movement of groundwater.

162

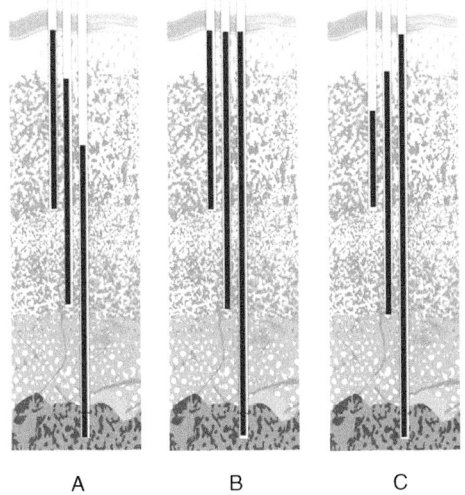

A   B   C

Fig. 5.22.

3. Groundwater towards the surface has a greater total energy per unit weight than the water at depth. Hence there is a downward movement.

23. Flowing Artesian Bore

The illustration is a schematic of a flowing artesian bore.

(1) What does the word 'artesian' mean?

(2) Match the following labels with the placeholders on the illustration in Fig. 5.23

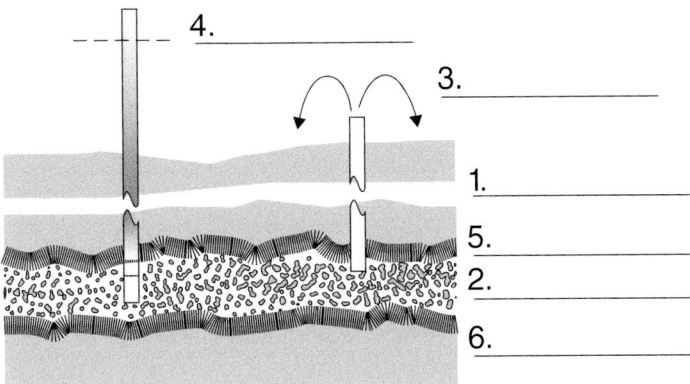

Fig. 5.23.

A. Piezometric pressure above ground level

B. Discharging bore

C. Vertical break indicating 'not to scale'

D. Upper confining layer (aquiclude)

E. Lower confining layer

F. Aquifer of permeable material

(3) Can you name aquifers with examples of flowing artesian bores?

24. Drainage Profiles

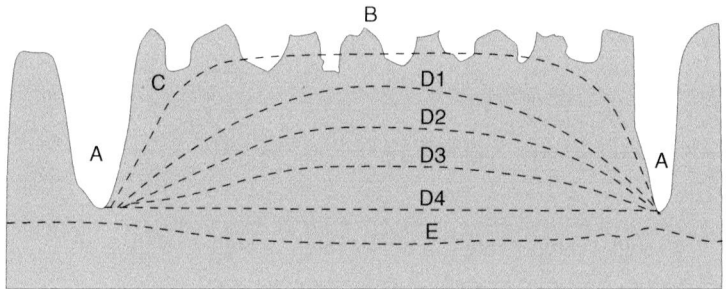

Fig. 5.24. Redrawn and based on profiles in Skaggs, R.W., Tian, S., Chescheir, G.M., Amatya, D.M. and Yousef, M.A. (2016) Forest drainage. In: *Forest Hydrology: Processes and Management*. CAB International, Wallingford, UK. The contribution of the authors is acknowledged.

The illustration shows a schematic (and vertically exaggerated) profile of a drained field with raised beds – this is the sort of arrangement that is often used in wet landscapes to grow trees or vegetables. Drainage is, in this case, by ditches but can often be by tile-drains. Link the letters on the illustration with the following labels:

1. Phreatic profiles at increasing times after rainfall.

2. Phreatic profile a long time after rainfall

3. Phreatic profile immediately after heavy rainfall

4. Drainage ditch (or tile drain)

5. Mounded beds

6. Phreatic profile without any rainfalls

25. Location of the Phreatic Surface

The illustration below indicates (in two dimensions) a common method of locating the phreatic surface relative to a water body by a line of bores. The stem of each bore has the height above the horizontal datum measured; for convenience this datum is set at 0 m for the stream water level. From the information given in Table 1 below, compute the phreatic surface gradient and angle for the three cases shown on the diagram in Table 2. Hint: complete Column 5 in Table 1, and then use information from Table 1 in Table 2. You will need to work out the gradient and then look these up in a tangent table.

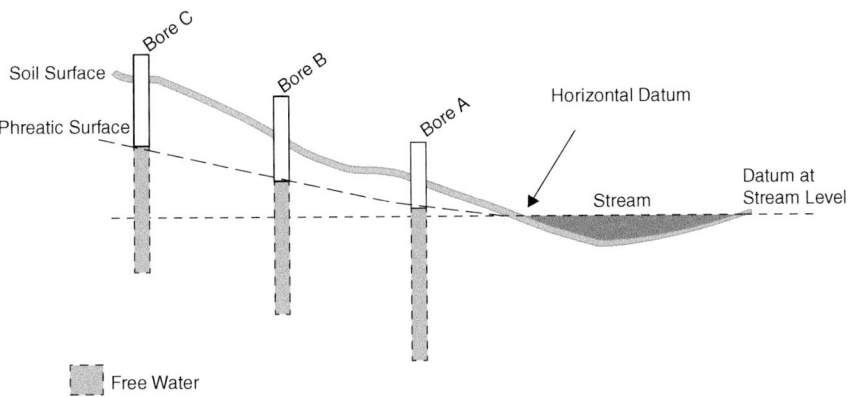

Fig. 5.25.

Table 1

Bore	Distance from stream, m	Height of stem above stream, m	Depth to water, m	Water surface level, m above stream
A	10	4.2	3.6	
B	30	8.6	6.2	
C	50	12.8	8.8	

Table 2

Reach	Horizontal distance, m	Groundwater rise, m	Groundwater gradient	Angle w/r to horizontal, °
Stream to A				
A to B				
B to C				

26. The Concept of Bank Storage

Bank storage occurs when a river bank also acts as an aquifer and the river has substantial variations in level associated with upstream flows. Look at the illustrations in Fig. 5.26 illustrating aspects of this process and then answer the questions. Note that time and height are best viewed as unitless and the illustrations are schematic, although these are based on the solutions of Cooper and Rorabough (1963). The upper illustration is a hydrograph. The lower illustrations show movement of water into the right bank during the rising and falling stages and some time after the flood pulse has passed.

Questions

1. Only the right bank is shown; what happens on the left bank?

2. For such a pulse, would the extent of the right bank matter?

3. If the river water held at a steady high level for a long, long time, would the answer to the above be the same?

4. Rivers with bank storage often suffer from bank failure. Why? What steps might be taken to prevent this?

River Hydrograph

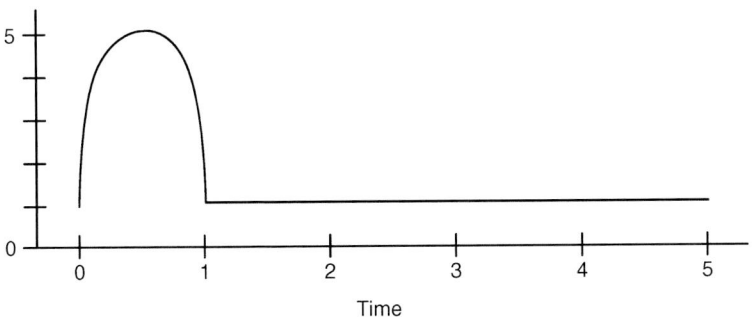

River/Bank Phreatic Profiles

Rising Stage

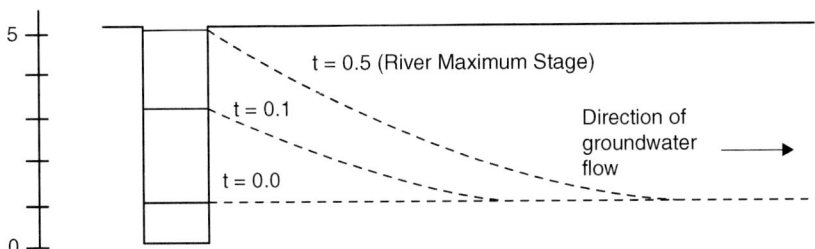

Falling Stage

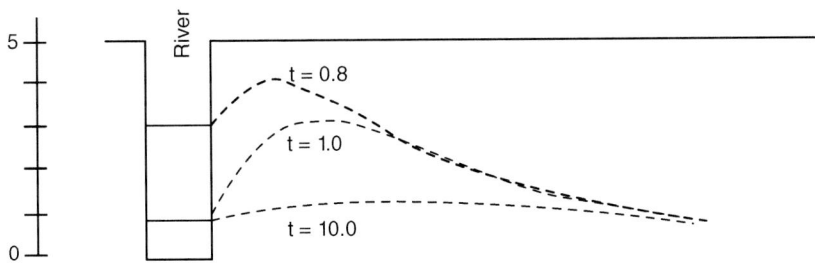

Fig. 5.26.

27. Evaporation vs Glacier-melt Diurnal Variations

When the sun shines, vegetation transpires and glaciers melt. Either can induce a diurnal variation in streamflow (but of opposite sign). Mutzner *et al.* (2015) partitioned these in a watershed in Switzerland and the results are shown for 3 days in the early autumn below. Answer the following questions:

A. Which hydrograph is from evaporation?

B. Which hydrograph is from glacial melt?

C. What does the * on the left lower side of each hydrograph represent?

D. If you have a watershed with both glaciers and vegetation, what happens on a sunny day?

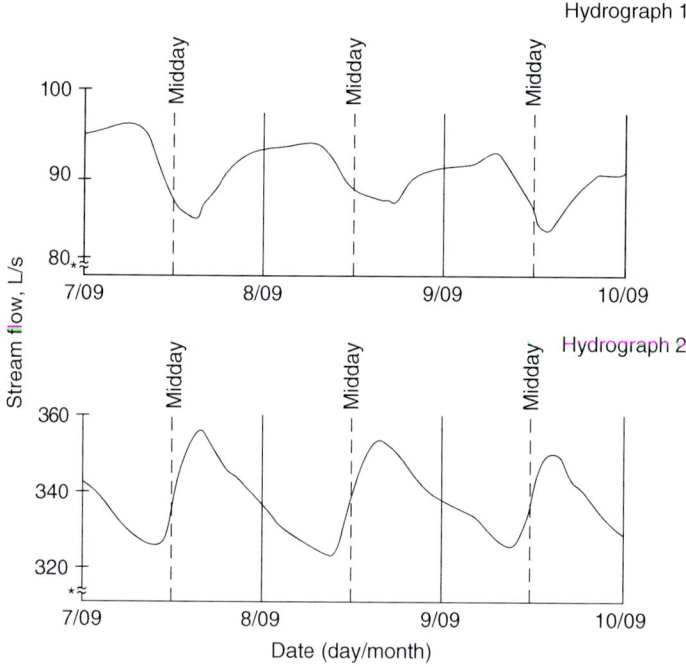

Fig. 5.27. Permission to redraw the illustration granted by author Raphael Mutzner.

28. A Two-colour Hydrologic World

The map below shows the world in two colours. Three possible interpretations are given. Which is correct (students should be able to find counter-examples to the incorrect two interpretations from their general knowledge)?

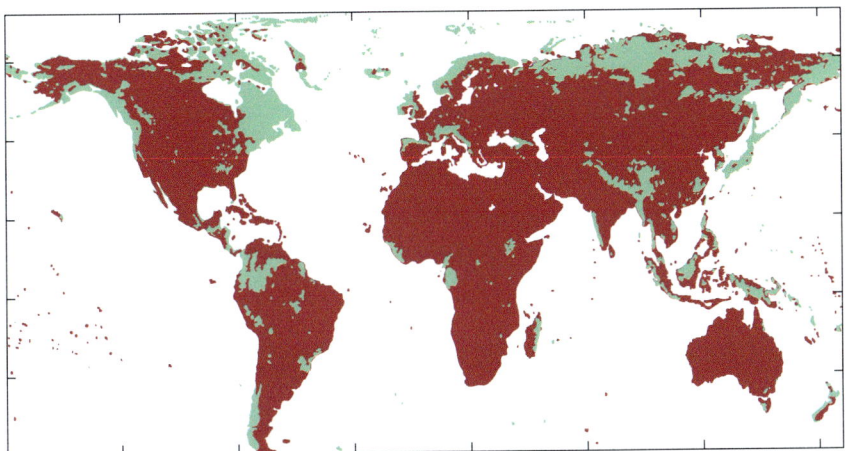

Fig. 5.28. Permission granted by Shaun Harrigan of the European Centre for Medium Range Weather Forecasts.

A. Areas in which demand for water by growing populations is predicted to exceed available water by 2040. These areas are shaded in red.

B. The mean annual discharge rate normalised by the mean annual precipitation rate (Q/P). Areas shaded in red are where evaporation exceeds runoff (i.e. Q/P<0.5, and E/P>0.5).

C. Areas in which effectively the available surface water resources are currently almost fully utilised, and which are predicted to come under severe stress from climate change by 2040. These areas are shaded in red.

29. Is This the Future of the World's Water?

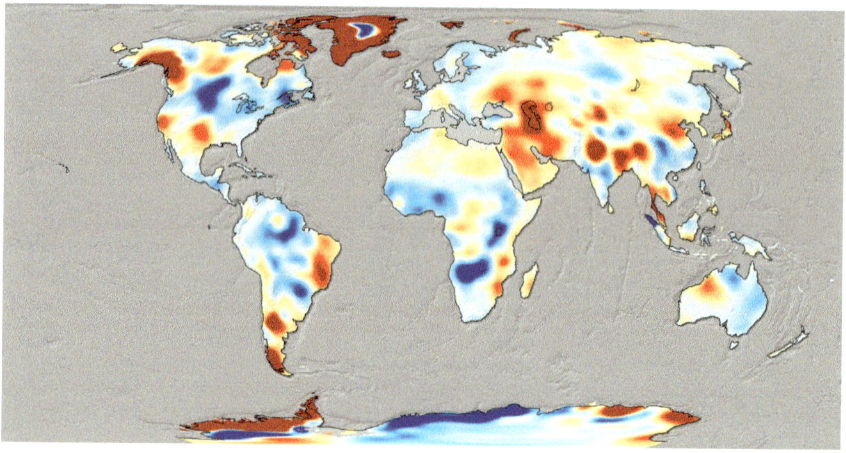

Fig. 5.29.

The map above was produced by the Pew Trust and was accumulated using 14 years of satellite data by the GRACE ('Gravity Recovery and Climate Experiment'). The data covers the period from 2002 to 2017. We have given five possible variations of what this map might be. Use your knowledge of hydrology to accept one and reject the other four propositions below.

A. Changed depth to fresh groundwater. Red means that the groundwater depth has decreased by 5 m, blue means that the available groundwater is 5 m closer to the surface than before.

B. The rate at which areas on the earth's surface are gaining or losing water. The dark red means that areas shown were 'drying out' by 2 cm year^{-1}. The dark blue shows that they were wetting up by 2 cm year^{-1}.

C. Change in forest evaporation rates over the last 17 years. Red means a greater transpiration (up to 100 mm year^{-1} increase), blue means reduced transpiration (100 mm year^{-1} reduction).

D. Change in global radiation impinging on the earth's surface. Red means that the areas are getting more radiation, and blue means that they are getting less radiation. The higher radiation received at the poles of the earth are striking.

E. Changes in the NDVI of vegetation in the area. The red means that the vegetation has diminished in health over the period of measurement. Blue means that the health of the vegetation has increased.

In addition, what is the probable map projection used? What are the advantages and disadvantages of this?

30. The Shading Shows Hydrologic Treasures – But What Treasure?

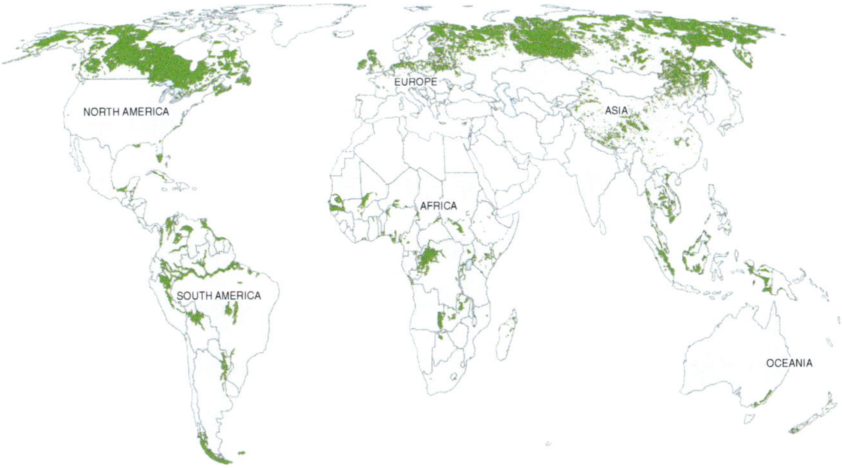

Fig. 5.30. Provided by GRID-Arendal (Cartographer Levi Westervold). Their contribution is acknowledged.

Yet another view of the world – this time in three colours. Which is the correct interpretation below?

A. The map shows the world's areas of 'very high rainfall' (>3000 mm per annum) shown as dark-green. The countries marked in grey have strong federal level control of water resources.

B. The dark-green areas of the map shows the world's areas where there are no major dams, and hence streams are 'natural'. The countries marked in grey have clear policies on protection of major rivers. Uncoloured areas appear to have no policies on dams.

C. The dark-green areas on the map show the global distribution of peat-lands – basically semi-permanent wetlands. Peatlands are known to occur in the countries marked grey but have not been fully mapped. These are important storages of carbon.

31. 'Zhang Curves', Runoff, and Watershed Efficiency

Zhang *et al.* (2001) derived a set of runoff curves using an international data set to show the mean evapotranspiration from pasture and forest landscapes as a function of mean annual rainfall. These curves have become embedded in the hydrologic literature and are colloquially known as 'Zhang Curves'. Fig. 5.31A shows the original form of curve which was derived from paired watershed experimental data analysed using a Budyko-based formulation.

171

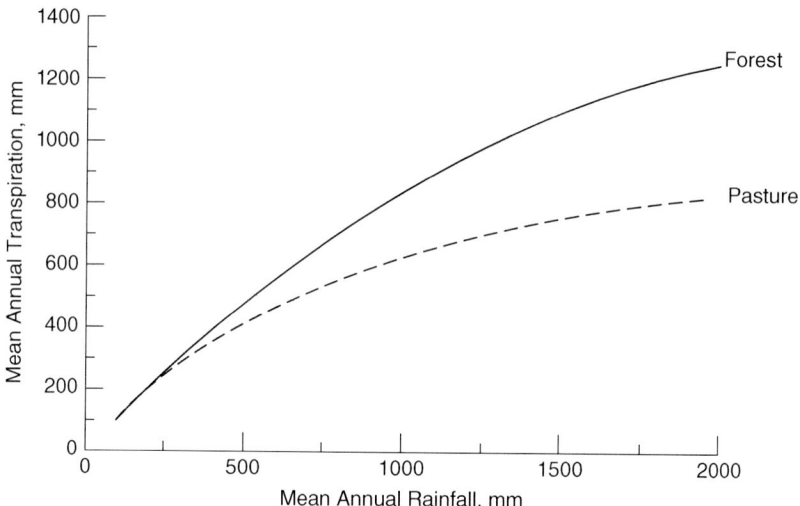

Fig. 5.31A. The usual form of 'Zhang Curves'. Courtesy of Springer Nature.

A. What is a 'runoff curve'?

B. The curves as shown go from 100 mm of rainfall, but the data comes from areas generally above 500 mm rainfall average. Is this a problem?

C. If you are given an annual rainfall, is it valid to estimate the annual ET using the curves?

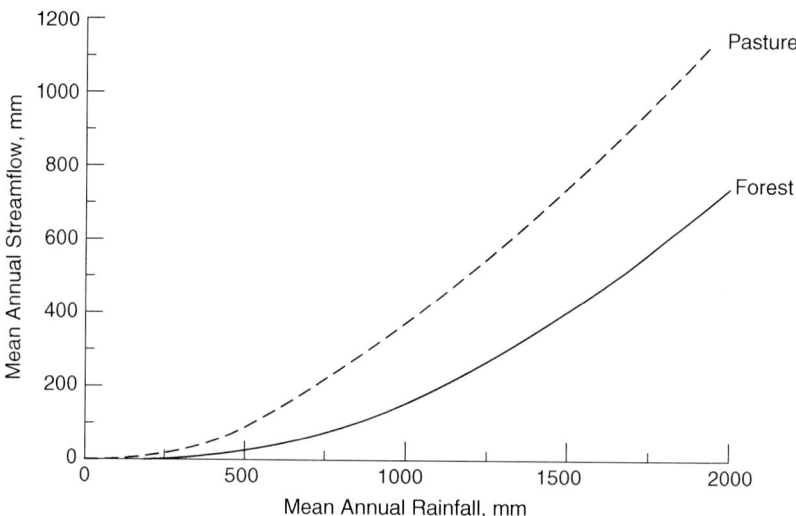

Fig. 5.31B. 'Zhang Curves' showing Mean Annual Streamflow as a function of Mean Annual Rainfall. Courtesy of Springer Nature.

D. Fig. 5.31B uses the curve in a different way and shows mean annual streamflow (P – ET) for forest and pasture as a function of P. Fig. 5.31C shows the mean watershed yield as a percentage of the mean annual rainfall. These rearrangements add to the ease of interpretation of these models. How do we go from Fig. 5.31B to Fig. 5.31C?

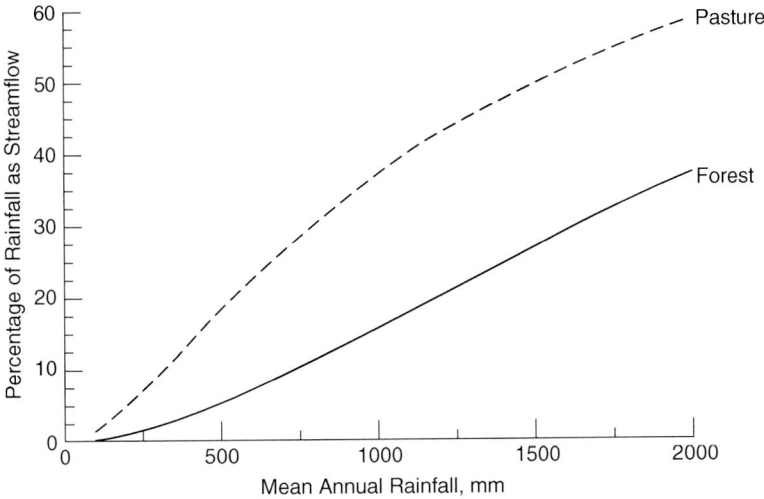

Fig. 5.31C. 'Zhang Curves' showing Percentage Runoff as a function of Mean Annual Rainfall. Courtesy of Springer Nature.

E. The curves show a significantly reduced water yield (or higher ET) from forests compared to pasture for a given mean rainfall. Why, then, are forests often the preferred land use for municipal watersheds when they use so much more water?

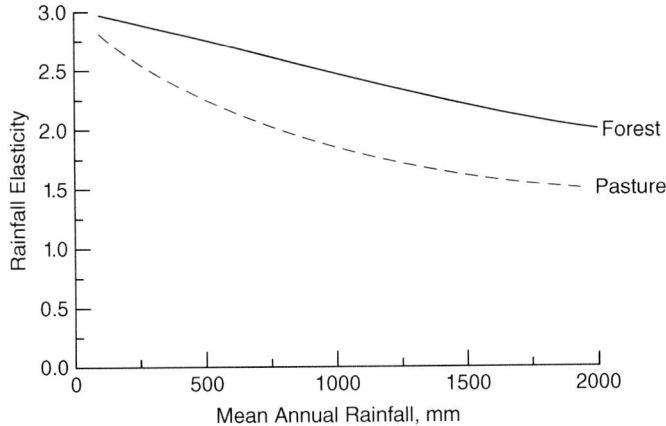

Fig. 5.31D. Rainfall elasticity as a function of Mean Annual Rainfall for both forest and pasture. Courtesy of Springer Nature.

F. Fig. 5.31D uses the curve of Fig. 5.31B to show the 'Rainfall Elasticity of Runoff'. This is the percentage change in streamflow for a 1% change in rainfall. Thus if the mean rainfall is 1000 mm a 1% reduction in mean annual rainfall would lead to about a 2% reduction in streamflow from pasture and 2.5% from forested land. Could you show this directly using measured rainfall and runoff data? Is this a useful measure?

How has the elasticity shown here been computed?

32. About Hydrologic Modelling

The schematic below illustrates the concept of hydrologic modelling. Look at this and answer the following questions

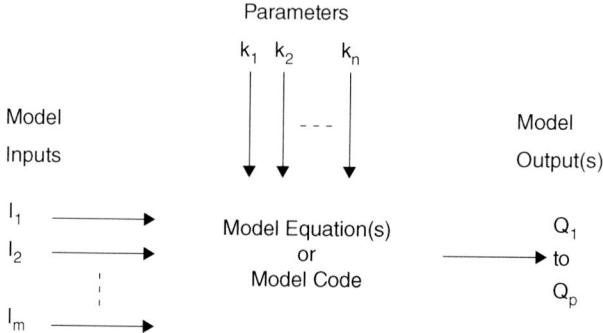

Fig. 5.32.

A. What is the distinction between a model input and a model parameter?

B. Are there 'grey' areas between inputs and parameters?

C. In the table below, classify the quantity given as an input, a parameter, or an output. The model is a streamflow model. Note that one quantity does not fit the classification.

Quantity	Classification
Hourly rainfall	_____
Channel roughness	_____
Crop albedo	_____
Estimated streamflow at 15-minute intervals	_____
Measured streamflow at 15-minute intervals	_____
Hydraulic conductivity of watershed slopes	_____

Continued

Continued

Quantity	Classification
Vapour pressure at 15-minute intervals	_____
Shape of watershed	_____
Recession coefficient	_____

D. We make a 'sensitivity test' by varying each of the parameters by 10% and seeing how the estimated streamflow varies. There is no change when we vary parameter k_4. Suggest interpretations for this.

E. How might we compare the computed outflow and the measured outflow?

F. We have physically measured parameter k_2 as having a value of 0.8. However, the measured sequence and the computed sequence really agree well when the parameter is set to 6, although its usual range is 0 to 1. What should we do?

G. The figure uses the subscript ranges 1-m, 1-n, and 1-p for inputs, parameters, and outputs, respectively. Is there or should there be any relation between the values of m, n, and p?

H. Suppose our watershed is broken into four sub-watersheds. How would this affect the inputs, parameters, and accuracy of the model?

33. Optimisation of a Physically Based Watershed Model (1)

Consider the model of the previous question. We have now divided the model parameters into 'measured parameters' (measured in the field), 'fixed parameters' (values taken from textbooks or papers), and 'optimising parameters' which we will use to get a good model fit. We have a measured sequence of output data for use in 'model calibration'. We can compare the fit of our model with this 'real' data sequence by computing the Nash–Sutcliffe (N-S) Coefficient of Efficiency in which a value of 1 means a perfect fit, and a 0 (zero) means that our model is performing no better than just using an average value would.

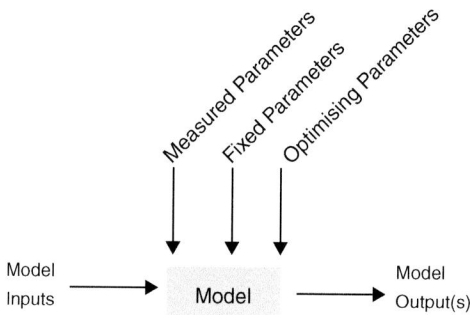

Fig. 5.33A.

We then set up a schedule shown below so that one or more 'optimising parameters' are systematically varied until we get the best possible measure of fit according to the Nash-Sutscliffe coefficient. In this, a set of parameters is fed in, the model run, the N-S coefficient compared to expectations, and the parameters either accepted or varied. The model is then rerun with the new parameters, and so on, until the best possible fit (highest value of the N-S coefficient) is obtained. The parameter variation is usually quite systematic.

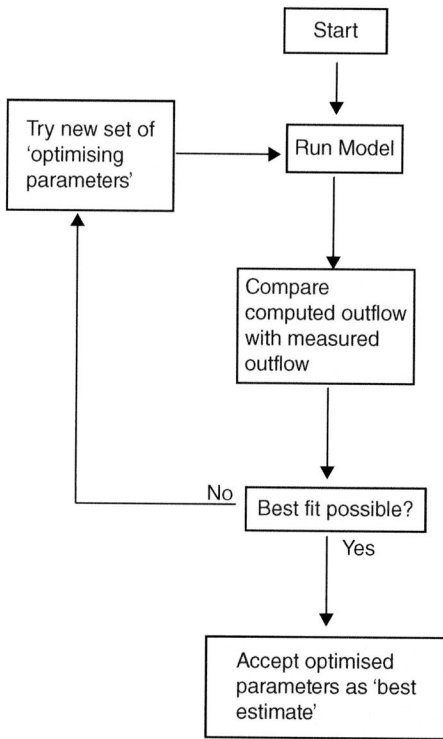

Fig. 5.33B.

A. Suppose we have three 'optimising parameters', k_1 to k_3. We have run our model with these against the calibration data and achieved an excellent fit. Does this mean that we have, de facto, determined the actual 'best values' of k_1 to k_3?

B. We have made a distinction between 'measured parameters', 'fixed parameters' (which we have set at some value and left unchanged), and 'optimising parameters' which we use to make the model more or less agree with the data. True or false – it's a pretty arbitrary distinction isn't it?

C. Elsewhere in this book, a distinction is made between 'physical models' that reproduce processes and 'black-box' models that produce an answer with no pretence that the actual processes are taken into account. Doesn't this procedure turn a physical model into a black-box model?

D. Suppose we had two optimising parameters, k_1 and k_2. Each is meant to range from 0 to 1. We make k_1 range from 0 to 1 in 10 steps. We do the same for k_2, giving us 100 (k_1, k_2) pairs. We then compute our N-S coefficient for each pair and plot the results in a 3D plot. This is called a 'response surface'. If it is a well-behaved model, what should the response surface look like? Early modellers using this approach encountered a number of difficulties; can you suggest what they might be?

E. Why were 'If' statements found to be diabolical in computing response surfaces?

F. Question D used the N-S Coefficient value as a function of (k_1, k_2) to give effectively a 3D model of a 'response surface'. Suppose we had three parameters (k_1, k_2, k_3). Would this approach still work?

G. It sounds like a lot of work – does the modeller have to sit there rerunning the model and noting the results? There must be better ways to do this?

H. Does this procedure bear any relation to the classic methodology of science?

34. Optimisation of a Physically Based Watershed Model (2)

The previous question talked about generating the Nash–Sutcliffe Coefficient of Efficiency as a function of two parameters, k_1 and k_2, to get our watershed model working as well as we can. The N-S coefficient was generated by comparing the model output with a real data sequence. Analogies were made to the surface generated on a (k_1, k_2) grid as being similar to a land-surface. Believe it or not, it is a good analogy. The principles apply to higher dimension cases (i.e. optimisation in three or more parameters), although the visualisation we are using here is not an option.

Fig. 5.34.

Let's take this further. Above is a photograph of Mt Buninyong (Victoria). We have drawn in three axes (k_1, k_2, Coefficient of Efficiency) and will pretend that this is a response surface. The point marked $(k_1{}^*, k_2{}^*)$ is a fire tower right at the top of the hill. Continuing on this theme:

A. If you were optimising on a two-parameter model, would this be a good response surface?

B. Would the best values of k_1 and k_2 for the model be those marked $(k_1{}^*, k_2{}^*)$?

C. True or false? The area in the foreground of the photograph could be called 'The Plain of Insensitivity' (or, perhaps, 'The Plain of Model Despair'). Why is this?

D. The algorithm we suggested was gridding the (k_1, k_2) space, running the model for all cases, and finding the best value. This works but is computationally inefficient and very time-consuming. Suppose we had an algorithm that arbitrarily picked a (k_1, k_2) pair, worked out which way was 'uphill', and then continually went 'uphill' until it could go no further. This is analogous to you getting onto the mountain and then always walking uphill to find the top. As you neared the top you might take smaller and smaller steps. Would that work in this case?

E. The photograph shows 'roughness elements' (trees, rocks, etc.). Would a real model response surface have anything analogous to these?

F. Under what conditions would such an approach to optimisation not work? What might be done to ensure this is does not happen?

G. Look at some of your local landscapes and view them as two-parameter response surfaces. Pretend that you were in a fog so that you could only see a few metres. Your task is to find the highest peak with this restricted visibility and no other knowledge. What difficulties might you encounter? How might your automated system avoid such problems? Remember that you are in a fog, so you can't look around much, but you do have an excellent memory for what you have seen.

35. A Budyko-Hydrology World

The Budyko curve is shown below. On the x-axis is the ratio of potential evaporation (Ep) and precipitation (P). The y-axis is the ratio of actual evaporation (E) and P. These are mean annual values. The dashed curve is a line representing over 1200 global watersheds analysed by Budyko (with an R^2 of 0.9).

A. If we know the ratio of Ep/P at a particular watershed is 0.8, what would we expect the ratio of E/P be?

A. About 0.5

B. About 0.7

C. About 0.2

D. There is not enough information

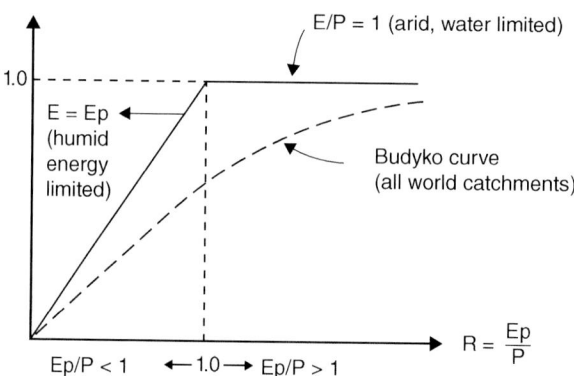

Fig. 5.35.

B. If a watershed plots to the right of Ep/P = 1, this means

    A. Potential evaporation is greater than precipitation, hence there will never be streamflow

    B. Potential evaporation is greater than precipitation, hence there will be always be streamflow

    C. There must be a lot of vegetation to evapotranspire

    D. The system, on average, is water limited

C. If a watershed had an E/P of 0.75, and Ep/P of 0.9, how much streamflow could we expect?

    A. 10% of precipitation

    B. 15% of precipitation

    C. 25% of precipitation

    D. I would need more information

36. Sodic (and Other ic) Soils

The structural stability of soils containing greater than 10% clay is largely dependent on the balance between calcium, magnesium and sodium ions in the soil solution. The photographs in Fig. 5.36A and B show erosion on a soil with an unusually high level of magnesium in Australia. Given that a sodium-dominated soil is called 'sodic', this may well be called 'magnesic'.

Fig. 5.36A. Photograph published with permission of photographer Terry Clark of Topo Group Pty Ltd, Queensland.

Fig. 5.36B. Photograph published with permission of photographer Terry Clark of Topo Group Pty Ltd, Queensland.

Questions

A. Are 'magnesic' soils unusual?

B. In either case, what leads to erosion?

C. Is soil piping ('tunnel erosion') present and what is the mechanism for this?

D. What is the impact on water supplies?

E. What treatments are available for sodic/magnesic soils?

F. What is the relationship of these to Question 16 ('What Is This?')?

G. Are these a worry in forested landscapes?

H. Are these only a worry for Australians?

37. Dryland Salinity

Dryland salinity is an issue in many countries, and it has been suggested was the cause of collapse in early civilisations around the Euphrates River. It is caused by removal of deep-rooted native vegetation and consequent hydrologic disequilibrium whereby a rising water table brings salts to the upper soil horizons. This is depicted below.

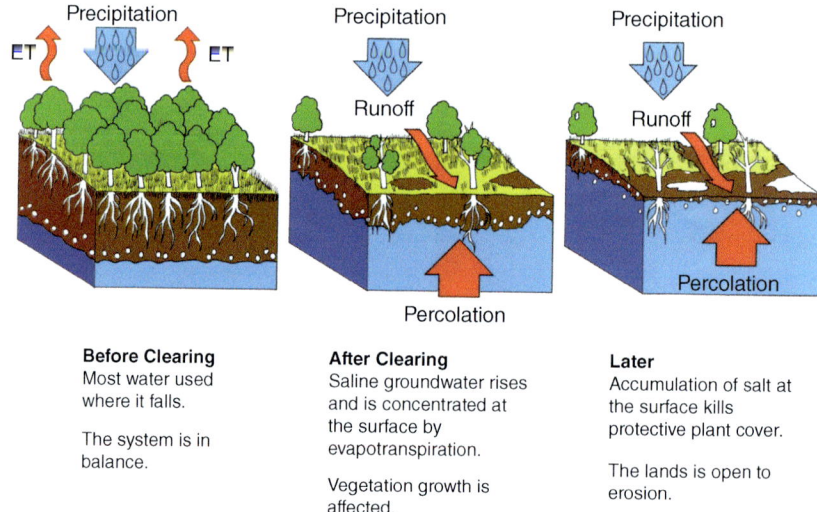

Before Clearing
Most water used where it falls.

The system is in balance.

After Clearing
Saline groundwater rises and is concentrated at the surface by evapotranspiration.

Vegetation growth is affected.

Later
Accumulation of salt at the surface kills protective plant cover.

The lands is open to erosion.

Fig. 5.37. Courtesy of Agriculture Victoria under their Creative Commons Attribution License.

Questions

A. What are the hydrologic processes that are in disequilibrium?

B. Salinity is often categorised by the terms below. Which category describes dryland salinity?

    A. primary salinity

    B. secondary salinity

    C. natural salinity

    D. irrigation salinity

C. Water tables do not need to rise to the surface to cause salinity problems. Why?

D. Mitigation of dryland salinity in agricultural landscapes has included the biological approach (as opposed to an engineering approach such as tile drains) of replanting deep-rooted vegetation, especially trees. If the aim is to reverse groundwater rise, which approach would be most likely to work:

    A. Planting in groundwater discharge zones?

    B. Planting in saline areas?

    C. Planting in groundwater recharge zones?

    D. Planting all over the area at high density?

E. Dryland salinity has some similarities to the 'Dust Bowl' events in the mid-western USA in the 1930s. What is the common thread?

Suggestions for Students
Tease out the relationship between irrigation and salinity in your local area. Find out about the history of salinity in the Tigris and Euphrates Valley. Compare the ancient salinity problems with the modern ones – will we suffer the same fate?

Answers

1. Heavy Rains

A. Both axes use logarithmic scales. The type of graph is called a Log-Log plot. This makes the data form an approximate straight line and fit neatly on the illustration.

B. Consider 400 mm hour^{-1} rainfall intensity. Now 1 ha = 10,000 m^2 and 1 mm depth = 1 litre per m^2. Also 1 litres weighs 1 kg. Hence this is 400 x 10,000 kg hour^{-1} being added = 4000 tonnes ha^{-1} hour^{-1} being added to the earth's surface. That's quite a loading and the weight alone is enough to induce landslides, cause roofs to collapse, and induce a range of 'unusual' (and usually detrimental) responses.

C. Sohra holds many such records (old and new) and is viewed as one of the wettest places on earth. It is mountainous and this interacts with the summer monsoon. However, it does suffer drought in winter. It is proud of its status as 'the world's wettest place' although more recently that title has moved to a nearby location.

2. Raster Hydrographs

A. Usually red is chosen as the most 'significant' or 'dangerous' colour. Response of people does depend on the choice of colours to some extent. It has been well shown that increasing the intensity of a restricted colour range communicates better than a full range of colours.

B. The choice of a logarithmic scale suppresses information about higher flows but does show up low-flow change. Depending on your interests, such highlighting or suppression may be an advantage or a disadvantage. Judicious choice of the colour scale will have a significant influence on the interpretation of data. One might have a few such plots with high flow, medium flows, and low flows 'spotlighted'. It is suggested that students 'play' with transforms that highlight some aspects of data and suppress other aspects.

C. Forming the plot into a cylinder about its vertical axis with December passing into January would make a truer representation. However, there would not be much information gain.

D. After about 1961 there is a marked reduction in low flows in the July to December period. This might be attributed to a change in the gauging station or methodology, a change in dam management policies, or possibly some sort of climate change effect. The first is most likely.

E. The plot could be displayed as some sort of contour map (large loss of information) or a 3D 'topographic' display.

3. Boundary of the Watershed

A.

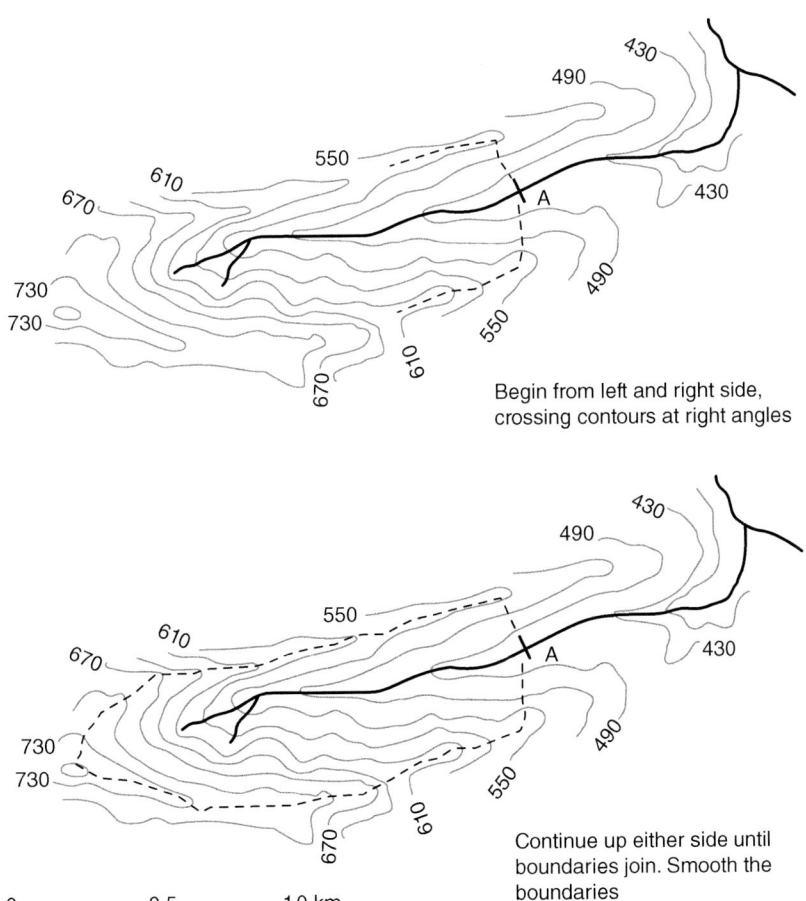

Begin from left and right side, crossing contours at right angles

Continue up either side until boundaries join. Smooth the boundaries

185

B.

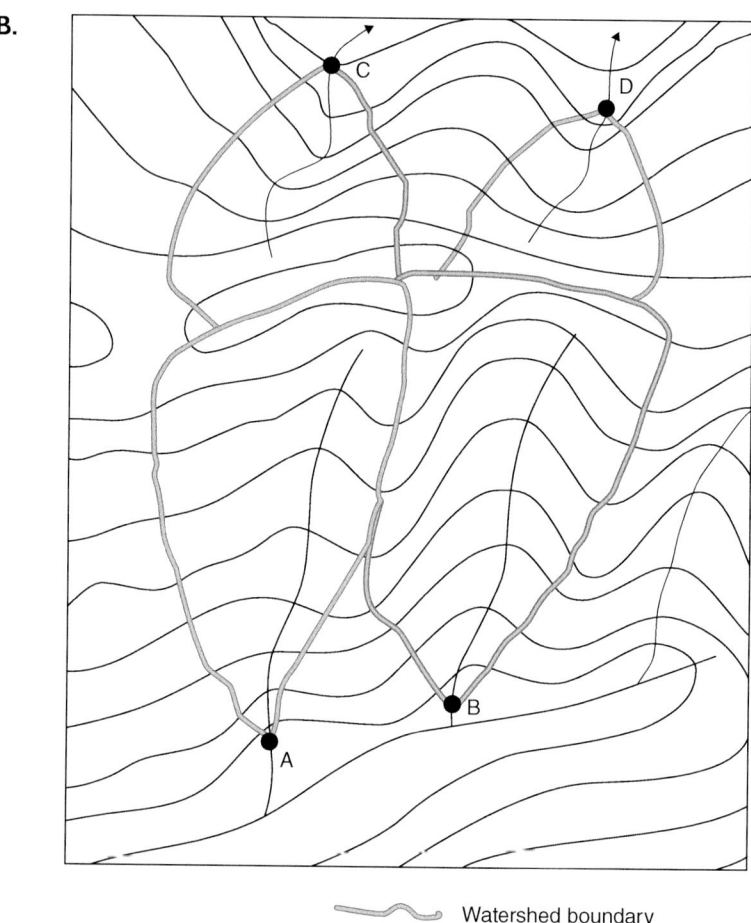

Watershed boundary

The streams should cross the contours at right angles, forming an 'orthogonal flow net' with the contours. This is a common issue with contour maps.

4. Watershed Boundaries on a Real Contour Map

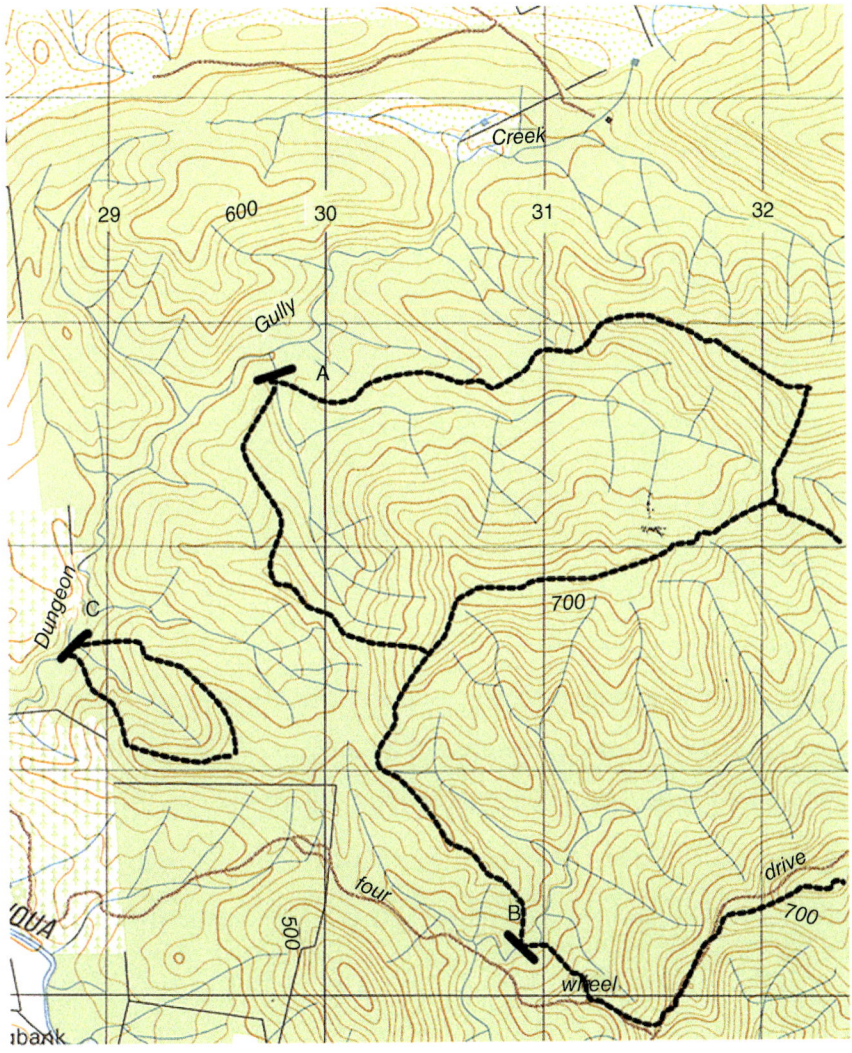

5. Strahler Ordering (1)

A.

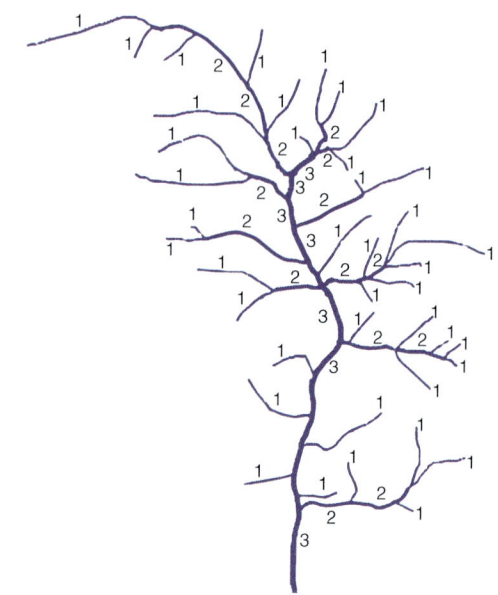

B. Strahler Ordering (2)

B. Rule 1, third-order

C. Rule 1, third-order

D. Rule 2, fourth-order

E. Ambiguous – could be 2nd + 2nd -> 3rd and 3rd + 3rd ->4th or successive applications of Rule 1 to give third-order.

F. Rules 1 and 2, fourth-order.

6. Big Floods and Bigger Floods

A. A 'Geological Event' would be something along the lines of a large landslide causing a lake to form. Sometime later the landslide mass fails suddenly, and a flood sweeps down the river. Contemporary examples were the 1980 flood in the Cowlitz River in Washington State, USA caused by the St Helens Volcano and the catastrophic flood due to the overtopping of the Vaiont Dam in Italy by a landslide in 1967. The

'Meteorological Events' floods are heavy rainfall, possibly onto snow. Sometimes these were consecutive events extending over one long period and large areas of land (e.g. Mississippi Floods of 1927).

B. No mistake here – it was a series of cataclysmic glacial lake outburst floods that swept periodically down the Columbia River Gorge at the end of the last ice age. They were the results of periodic ruptures of ice dams. A cycle of flooding and reformation of the lake lasted an average of 55 years. The floods occurred a number of times over the 2000 year period between 15,000 years and 13,000 years ago. There is evidence of at least 25 massive floods with the largest discharging about 10 km^3 $hour^{-1}$ of water. It is speculated that the flow speed approached 130 km $hour^{-1}$ (an almost unimaginable velocity for surface water and one that no land surface could withstand). This flood was one of a number of huge floods of the past occurring over the earth. Students should look up the 'Bonneville Flood' too. One could ponder how we would cope with such an event today.

C. The USA is a large country with an unusually diverse range of landscapes and weather so it would be expected to have lots of 'interesting points'. However, it also has an excellent system of weather recording, a well-educated population, and the data is accessible to all. Hence in such studies it will feature prominently.

D. The usual experience is that people's perception of a flood as 'huge' reflects the impacts on their life and property rather than the actual volume of water. It is unusual to have deaths in such floodings (other than by people trying to drive through flood waters). Most such floods, plotted on this type of graph are at the lower end. However, the economic magnitude of such floods can still be large. Floods which actually are of large magnitude and pass through urban areas (e.g. the New Orleans Floods of 2005) are very destructive in social and economic terms.

E. There are not large land masses at higher latitudes in the southern hemisphere, and hence little or no flooding at these latitudes. The asymmetry of the flooding line in regard to the equator has no clear explanation, but factors such as the topography of the land masses near the equator would play a role.

F. The floods in the Yangtze River are very large and, given the high population density can be extremely costly in economic and human terms compared to the Australian flood experience. Although these areas experience large floods, the food production of these areas rely on abundant water resources and the heavy, monsoonal rainfall – perhaps it is a price they pay for an abundant and reliable rainfall? The Australian floods shown caused large economic disruption and widespread areas of shallow inundation but little loss of life. These floods did show the vulnerability of southern Australia to hydrologic disruption of economic activities.

G. There are no extensive land areas subject to snow in the 'high' southern latitudes, and hence no large-scale floods. Snow in the southern hemisphere is generally limited to high mountains. Snow falling at sea in the southern hemisphere rarely happens – probably reflecting the strong, turbulent winds and the heat stored in the ocean.

Reference: O'Conner, J.E., Grant, G.E. and Costa, J.E. (2002) The geology and geography of floods. In: *Ancient Floods, Modern Hazards: Principles and Applications of Paleoflood Hydrology. Water Science and Application.* American Geophysical Union, Washington, DC. Volume 5, pp. 359–385.

7. Impacts of Burning on Watersheds

Time of burning (1) is shown by label C. Streamflow momentarily decreases, the diurnal variation ceases, and flow increases.

The spike flow (2) is shown by label A. This is a short-lived but high peak and generates major flows.

The diurnal variation (3) is shown by label B. This is due to transpiration in the riparian zone.

The increase in flow (4) is shown by label D. This reflects a decrease in evapotranspiration due to the destruction of the tree canopies.

The data is from O'Loughlin, E.M., Cheney, N.P. and Burns, J. (1982) The Bushrangers experiment: Hydrological response of a eucalypt catchment to fire. In: *The First National Symposium on Forest Hydrology 1982.* Institution of Engineers Australia, Canberra.

8. Properties of Water

Both illustrations show the phenomenon of surface tension of water, and its ability to adhere to surfaces. This leads to capillarity. In general, the meniscus of the pipette is read at the lowest points. Early workers in soil physics compared water in soils to water in ever-finer capillary pores, although the analogy broke down because such capillary surfaces could not be demonstrated in actual soils. This view was displaced by Edgar

Buckingham in 1904 who came up with the current theories of soil water potential which have little in common with capillary theory. His model has stood the test of time but was bitterly resented by soil physicists who built their careers on capillary analogues.

9. Water to Ice

The ice cube is formed using 'heavy water' (deuterium) and placed in ordinary water; if it was in deuterium water it would float. Deuterium water does occur as an isotope in natural water but this situation would not occur in nature.

10. A Common Water Supply Issue

The illustration shows filming associated with high levels of dissolved calcium ('hard water') in the water. This usually comes from limestone geologies in the watershed and can be difficult to remove. Often the water has products based on sodium hexametaphosphate (generically known as 'Calgon' after a successful product of that name) added to prevent cleaning interference. Although dissolved calcium is a major cause of hard water, other ions (including iron and magnesium) are also often found.

11. Hydrometric Devices

A. Water level recorder. The float-counterweight system rotates a circular drum on which a chart is fixed. A clock drives a pen along a set of rails. The result is a record of water level over time. This is an early form of data logger. Although 'obsolete' they last a long time, are reliable, and give a very graphic record. See also Question 12 for an example of the output.

B. A stage-level measuring device for easy reading of water levels ('stage'). Typically this would be housed in a stilling well and read by an observer at a fixed time. Such records are now viewed as important for climate change analysis.

C. A neutron probe with integrated rate scaler. The red box is mainly a shield for the actual probe which is lowered into the soil. They are heavy to carry around slippery hillside slopes, difficult to calibrate, and it can take an hour to log a 6 m tube, but they work.

D. A standard 8" (203 mm) rain gauge in a stand. A daily-read rain gauge is the basis of so much of our hydrology knowledge.

E. A paddle-type current meter used to give a measure of flow velocity. For shallow streams the observer wades in, lowers the device, and reads the water velocity. The mean velocity is multiplied by the area of cross-section to get the flow.

F. A discrete water quality sampler. The mechanism purges lines, sucks up a sample at a preset time, stores it in a bottle and rotates the mechanism to the next bottle. These work well but can put a very heavy analytical load on laboratories. And the samples are heavy to carry around.

G. The image (courtesy of the US Geological Survey) shows a temporary weir being used to measure streamflow with a weir wall formed by mud.

H. Instrument standards for turbidity. These are a suspension of latex. In most forested watersheds the turbidity would be close to 5 or less.

12. How Things Used to Be Done

A. The paper chart is driven by a clock which is powered either by a spring or a falling weight. Typically the clock would need winding once a month. When well-maintained, they were very reliable.

B. The chart could be digitised by using such a photograph. Typically, four points of known time and height would be used for calibration. The cursor would then be moved along the line and a stream of time-height coordinates would come out. The program 'Didger' (Golden Software) is an example of software commonly used for this. Sometimes a large 'digitising table' is used to ease the handling of large charts and to improve accuracy and speed of the task.

C. Yes, it meets definitions of being a data logger, although it wouldn't be recognised as such by most younger scientists. However, the technology of data logging changes continuously so care is needed in defining what constitutes such a device.

D. Pen charts give a graphic record and failure of the recording system is very apparent (unlike more sophisticated systems). Many variations are much more apparent on charts than in digital records. The ability to write annotations on charts adds a lot to interpretation which may come many years or decades later. The charts require little storage other than a dry space. However, digitising can be tedious and prone to errors, and the logistics of ordered storage can be difficult.

13. Watershed Hydrology and Political Boundaries (1)

A. You are looking at a flood plain with the main course of the river (at low flow) meandering through it. Some past meanders have been 'cut-off' forming cut-off meanders (oxbow lakes, billabongs, anabranches, etc.). These are periodically reconnected with the river at periods of high flow. Ecologists note that a lot of biota that struggles to survive in the main channel survives in these 'refugia'.

B. The main course is that marked by the arrow and A.

C. The 'high bank' of the Murray would be a line running along the northern edge and southern edge of the meanders, but this has proven too difficult Thus the Victoria-NSW boundary is the southern bank of the main (low-flow) course. This means that the river is administered by NSW – thus you might be fishing on the Victorian bank but you will need a NSW fishing license. Pleasantly (and at the present) the river is not a 'wanderer' so the boundary has been stable for a century or more. The difficulty with the point marked A is that the Victorian boundary here is on the northern side of the river. There are three or four other such reaches in this sketch alone. Issues such as this show the difficulties of using a river as a political boundary.

D. It is likely that the sand will be 'dredged' from the river in the hope of avoiding an avulsion. The River Murray has been stable so such an avulsion would be a new exercise in Australian politics. Although the States and Federal Government have a 'natural river' policy, this would probably be viewed as a little 'too natural'. There would be pressure to do engineering works to return the flow to the pre-avulsion channel. Whether these would work would be anyone's guess.

14. Watershed Hydrology and Political Boundaries (2): The Gambia

A. The advantage is that nowhere in The Gambia is far from either their coast or the river and these are the major sources of wealth for the country. River transport can be an efficient means of travel within the country.

B. It means that Senegal loses access to the river and it makes north–south travel in Senegal difficult because one has to go around The Gambia. At various times the two countries have united but there seems to be no desire to have a union. It does make going from north to south Senegal about twice the straight-line distance.

At various times there have been major drives to have administrative subdivisions based on being within a given watershed. While such a division has certain advantages, the watershed boundary crosses many other boundaries. Thus there are no overwhelming advantages to this approach.

15. Erosional and Depositional Forms

A. Erosion of the mountains in the upland streams has deposited sediment in a downstream area, changing a V-shaped valley to a flat-bottomed valley. It's hard to believe that the original valley bottom was once on bare rock! The water is insufficient to cover the entire valley and has formed preferential courses – this is called 'braiding' and is characteristic in flat, alluvial rivers. Occasionally, forestry rehabilitation work on upland slopes has reduced the inflow of sediment to the valley floors, and this has resulted in erosion of these floors.

B. The photograph shows deposits of colluvium at a lower slope. This is upslope sediment that has moved downslope by landslides large and small and formed a mixed mass of soil, saprolite, and rock. Although water is usually involved, it is not an alluvial process, and the rocks show no sign of being water-worn. The presence of deep-rooted forests is usually viewed as one method of preventing this form of failure.

C. The photograph shows the deposition of a small alluvial fan at the outwash of a headwater stream a year or so after burning. Before burning the stream channel was V-shaped and on bedrock. After burning a miniature flood plain formed from soil washed into the stream. Some of this was deposited as an alluvial fan at a change in the stream gradient. Two years after this photograph the vegetation had covered this and there was no obvious sign of this geomorphic process. Small mountain streams are very dynamic.

D. A 'nick point' formed by a log in the same stream as C. This gave a point of stability in the bedrock channel and allowed sediment to accumulate. It has been shown that the flow of woody debris into small and large streams is an important factor in their natural stability.

E. A coastal landslip. Factors include weak weathered rocks, lack of deep-rooted biophysical stabilisation, heavy rains weakening the soil strength and increasing the weight. Small earthquakes can be potent initiators. The ocean plays a role too, by giving high levels of sodium that can disperse the clays and undercutting the cliffs, thereby removing the slope toe.

16. What Is This?

In fact it is B (tricked you?) but it could equally well be A or C. The network type is 'dendritic' and dendritic drainage is the model for most watersheds. This network is well described by 'fractal models' and occur across a range of scales, ranging from many kilometres across to a few millimetres across. The fractal concept implies that as you 'zoom' into it, the same level of detail becomes apparent across a wide range of scales. See if you can find local examples of dendritic networks (often eroding cliff-faces are a good spot to start).

17. Landscape Evolution

B is in the Canary Islands and would be 'Short storm/intermediate'. C is in the Canberra hills of Australia and is a good example of a 'long storm' landscape. D is from 'Badlands' in Utah and would be classified as 'short storm'. Notice the dendritic form (small scale) referred to in the previous exercise developing on the slopes. E is the coastal ranges of Oregon and is a 'long storm' landscape.

18. Groundwater/Stream Relationships (Influent and Effluent Streams)

A - 4, B - 5, C - 7, D -1, E - 3, F - 6, G -2 or (rearranged)
1 - D, 2 - G, 3 - E, 4 - A, 5 - B, 6 - F, 2 - G

19. Groundwater Pumping and Cones of Depression

A - 3, B - 2, C - 5, D - 1, E - 4, F - 7, G - 6

20. Coastal Aquifers (1)

A - 5, B - 9, C - 3, D - 8, E - 1, F - 7, G - 6, H - 4, I - 2.
The water in the piezometers would be fresh.

21. Coastal Aquifers (2)

A. Inland groundwater pumping has reduced the hydraulic head of fresh water. BC and BD have thus decreased. This decrease of freshwater pressure has allowed the level of salt water to rise.

B. The inlet of the bores is now in the salt-water zone and the bores would be producing salt water. This change can occur quite fast and can be toxic to crops being irrigated.

C. Strategies include developing knowledge of the interface behaviour, limiting pumping in the aquifer to avoid salt-water rise, and automatic salinity detection which shuts pumps down if water exceeds a threshold conductivity.

22. Interpretation of Multi-depth Piezometers

A - 3, B - 1, C - 2.

Excess energy is dissipated ultimately as heat.

The nest of piezometers gives us no information about the horizontal direction of movement. Usually a number of other nests would allow a 3D model of groundwater movement to be built up.

23. Flowing Artesian Bore

1. Artesian means a well bored into a pressurised aquifer. Thus, as in placeholder 4, the piezometric pressure is above ground level. Thus water will spontaneously flow out of a well without a valve.

2. A - 4, B - 3, C - 1, D - 5, E - 6, F - 2

3. Examples include many wells in the Great Artesian Basin of Australia and the Ogallala Aquifer in the USA. In both cases, excessive use of the groundwater is reducing artesian pressures.

24. Drainage Profiles

A is 4 (drainage ditch)

B is 5 (the mounded beds)

C is 3 (phreatic profile immediately after rainfall)

D1 to D3 are three phreatic profiles at increasing times after rainfall

D4 is 2 (the phreatic profile a long time after rainfall)

E is 6 (the phreatic profile in the absence of recent rainfalls). This often shows a dip in the middle of the mound because of ET of the vegetation.

25. Location of the Phreatic Surface

Table 1

Bore	Distance from stream, m	Height of stem above stream, m	Depth to water, m	Water surface level, m above stream
A	10	4.2	3.6	0.6
B	30	8.6	6.2	2.4
C	50	12.8	8.6	4.0

Table 2

Reach	Horizontal distance, m	Groundwater rise, m	Groundwater gradient	Angle w/r to horizontal, °
Stream to A	10	0.6[*1]	0.06[*2]	3.4°[*3]
A to B	20	1.8[*4]	0.09	5.1°
B to C	20	1.6	0.08	4.6°

[*1]. Water surface level above stream = 0.6 = 0.6 − 0 (i.e. g/w height − stream height)
[*2]. Gradient = 0.06 = 0.6/10
[*3]. Tan⁻¹(0.06) = 3.43°
[*4]. Groundwater rise = 1.8 = 2.4 − 0.6

26. The Concept of Bank Storage

1. It is usually assumed the left bank mirrors the right bank.

2. For a short-lived pulse, the water would usually only penetrate a short distance into the aquifer.

3. For a long-lived high flow water may penetrate many kilometres from the river and recharge substantial, interconnected aquifers.

4. Returning water creates high groundwater pressures inside the bank. These can be enough to 'blow' the bank outwards. Biophysical reinforcement of the bank is one solution. Managing the river so that rapid changes in river height are avoided is another common practice.

Reference: Cooper, H.H. and Rorabough, M.I. (1963) Groundwater movement and bank storage due to flood stages in surface streams. *US Geological Survey Water Supply Paper* 1536-J.

27. Evaporation vs Glacier-melt Diurnal Variations

A. Hydrograph 1 shows a decrease in flow in the afternoon. This is an evapotranspiration signature.

B. Hydrograph 2 shows an increase in flow the afternoon; this is a snow-melt signature.

In both cases there is a time-lag between when the maximum rate of either process occurs and when the maximum change in streamflow occurs.

C. The asterisk marks a break in the vertical axis ($\approx$ or ellipsis) to show that it does not start at zero.

D. On a sunny day where both are present the diurnal variations tend to cancel one another out, although usually the snowmelt signature is stronger.

Reference: Mutzner, R., Weijs, S.V., Tarolli, P., Calaf, M., Olroyd, H.J. and Parlange, M.B. (2015) Controls on the diurnal streamflow cycles in two subbasins of an alpine headwater catchment. *Water Resources Research* 51, 3404–3418.

28. A Two-colour Hydrologic World

Answer **B** is correct. The map is based on EU-WATCH data over the period 1901–2000 and is available on the Internet.

From Harrigan, S. and Berghuijs, W. (2016) The Mystery of Evaporation. *Streams of Thought (Young Hydrologic Society)*, published July 2016. DOI: 10.5281/zenodo.57847

29. Is This the Future of the World's Water?

Answer **B** is correct. The others are all figments of the author's imagination. The state of the polar regions provide a big clue to their falsity (no trees, no groundwater). The map shows the rate at which areas on the earth's surface were gaining or losing water over the period from 2002 to 2017.

GRACE, which was launched in 2002 and decommissioned at the end of 2017, was close to a 'scale in the sky'. It measured the very tiny space-time variations in earth's gravity field, effectively weighing changes in water mass over large river basins and groundwater aquifers–those porous, subterranean rock and soil layers that store water that must be pumped to the surface.

In particular, the wetter areas appear to be getting wetter and the dryer areas are getting dryer. The map has caused considerable conjecture about the future of the world's water and the relation to climate change. The author (Professor Famiglietti) concluded that water security – a phrase that simply means having access to sufficient quantities of safe water for our daily lives – is at greater risk than most people realise.

The map was produced by Professor Jay Famiglietti. He is a professor and the executive director of the Global Institute for Water Security at the University of Saskatchewan, where he holds the Canada 150 research chair in hydrology and remote sensing. See the discussion available at: A Map of the Future of Water | The Pew Charitable Trusts (pewtrusts.org)

Although the map projection is not stated it is similar to the Mercator's Projection (once commonly used but now viewed as AngloCentric). This works well for the middle latitudes of the earth but gives gross distortions at higher latitudes.

30. The Green Shows Hydrologic Treasures - But What Treasure?

Answer **C**. Appreciation of peatlands has only come recently and are now viewed as amongst the world's underappreciated hydrologic treasures. Peatlands play a large role in water storage in watersheds, store carbon, and give unique landscapes. Around the world many areas of peatland that were converted to agriculture or commercial forestry sites are being reconverted to peatland. This also partly reflects their low nutrient status and difficult site conditions for commercial enterprises. It is suggested that students visit some local peatlands to get a feel for these.

The map was drawn in 2017 by Cartographer Levi Westerveld, and is from 'Smoke on Water' Collection. See website Distribution of global peatlands | GRID-Arendal

31. 'Zhang Curves', Runoff, and Watershed Efficiency

A. A runoff curve is usually the streamflow as a function of one or two variables (usually including rainfall). They are popular because they give a good overall representation.

B. At low rainfalls, pasture is doubtful and forests in the usual form do not exist. In most countries there is very little runoff below about 400 mm mean rainfall. Hence to some extent the low rainfall is an extrapolation.

C. The input is mean annual rainfall so an annual rainfall by itself is not a valid input. This stricture is often ignored. It can be shown that although the mean of locations fits quite well, the change of annual runoff with a change in annual rainfall may not always be in accord with the mean curves. In particular, runoff at particular sites can diminish dramatically at low rainfalls.

D. Fig. 5.31B is (P − ET) for forests. Fig. 5.31C is 100 (P − ET)/P. Of particular interest is how the runoff increases substantially with rainfall, which is why high rainfall areas are the preferred location for municipal watersheds.

E. Forests tend to allow infiltration of rainfall, and this gives a much steadier outflow than pasture. This is suited to the needs of municipal reservoirs. In addition, the water is usually cleaner and with less bacterial contamination, and thus needs less treatment. Thus, although the yield is less, forests tend to be the preferred watershed land use for municipal watersheds.

F. Rainfall elasticity is a useful thing for scientists to know about. It would be very hard to contemplate and compute rainfall elasticity without using runoff curves. At very low mean annual rainfalls, this 'elasticity' becomes marked but because the runoff is so low, it is not really a valid measure. Occasionally, in drought-stricken societies, the sensitivity of streamflow to rainfall causes a feeling that 'Mother Nature is against us all'. This is exacerbated by the fact that a decline in rainfall will cause a greater percentage decline in runoff than the corresponding increase in percentage runoff for an increase in rainfall.

The elasticity shown here is the first derivative of the runoff with respect to rainfall and was computed directly from the analytical expression for these curves. This can be done on paper using the 'Chain Rule' but in this case was done more simply using the Wolfram package 'Mathematica'. It could also be done graphically or numerically.

From Zhang, L., Dawes, W. R. and Walker, G. R. (2001) Response of mean annual evapotranspiration to vegetation changes at catchment scale. *Water Resources Research* 37(3), 701–708.

32. About Hydrologic Modelling

A. A parameter is fixed over the course of the modelling and, in principle, can be measured independently. An input varies in time and space over the modelling run.

B. Some parameters can vary over the modelling run – for instance 'Manning's n' coefficient of channel roughness often varies with the flow in the channel. This makes a 'grey area'. Some models have parameters that change with the output but this makes for complex programming and gives discontinuities or 'unusual' results when the parameters change.

C.

Quantity	Classification
Hourly rainfall	Model input
Channel roughness	Parameter
Crop albedo	Parameter
Estimated streamflow at 15-minute intervals	Model output
Measured streamflow at 15-minute intervals	Not in classification
Hydraulic conductivity of watershed slopes	Parameter
Vapour pressure at 15-minute intervals	Model input
Shape of watershed	Parameter
Recession coefficient	Parameter

D. The model is insensitive to parameter k_4. This might mean that the model is in a range where this does not feature, the influence of the parameter is so small that it is undetectable at the numerical level of the output, or there is a mistake in the model code.

E. Model outputs are compared to 'real data' by plotting the two on a common time axis and looking carefully at it. A quantitative approach is computing parameters such as the Nash–Sutcliffe 'Coefficient of Efficiency'. This takes a value of 1 for a perfect fit, 0 for a fit no better than the mean, and a large negative value when it is really bad.

F. You've got a problem on your hands. You might try to see why the model behaves so. Ideally parameters should be independently measured but sometimes this is not possible. In some cases the parameter set that gives the best output would be viewed as 'optimised' but the parameters should always be within their correct range.

G. The values of m, n, and p are arbitrary and set by the model designer. However, the complexity of a model goes up as about the square (or the cube?) of any of these, so that they should, ideally, be kept as low as possible consistent with the aims of the model. A model where p exceeds m or n would be viewed with suspicion by hydrology researchers (i.e. a small amount of input leads to a large amount of output – the reverse of what science is meant to be about).

H. Division of the model into four separate domains would, ideally, require four separate input streams and four separate parameter sets. This may enhance the accuracy of the output but makes the model far more complex.

33. Optimisation of a Physically Based Watershed Model (1)

A. If the values of the other parameters are valid, and if the model has a physical structure that mimics the real physical structure then, yes, you have made a 'best estimate' of the true values. However, the two 'ifs' are big ones so this is unlikely.

B. True. The parameter subdivision is arbitrary, and it's 'sausage machine computation' that will churn out an answer irrespective of this subdivision. For better or worse, the results are often not very different over a wide range of parameters.

C. The optimisation process helps give a good model fit, and the parameters selected achieve this. The process usually reduces the advantages of the physically based model and makes it more like a 'black-box' model, but at least it appears to work.

D. The optimisation process relies on the assumption that a small change in a parameter should give a small change in outputs. The ideal output would be a smooth 'hill' with a clear maximum point (i.e. best (k_1, k_2) pair) and that falls away in all directions from that maximum point. Early researchers found that the 'response surface' they computed was not smooth but had many discontinuities ('cliffs') because of programming mistakes, 'If' statements in the computer code, underflow, and overflow errors. Sometimes there were multiple peaks in the (k_1, k_2) domain (i.e. competing pairs of 'best values'). Sometimes the surface was flat (i.e. the model was insensitive to the (k_1, k_2) values chosen. A common occurrence was that the 'highest point' was at the edge of the (k_1, k_2) domain. This suggests that a better result would be obtained by taking k_1 or k_2 outside their (0–1) range.

E. 'If' statements in computer code often put the model on a different path, depending on the value of something in the model. Thus, the statement If (a>0.5) Go To. . . would result in a discontinuity in the output from this point. This has proved particularly difficult in model optimisation.

F. As used here, the N-S coefficient as a function of two parameters can be visualised in 3D space. If we have three or more parameters, we can not visualise it in this way other than by taking 'two-parameter' slices of it. The mathematical principles are the same.

G. The computation of the response surface can be fully automated (see, for instance, the 'PEST' (Parameter ESTimation) suite of software). However, it is not uncommon for the modeller to run through all combinations.

The automated programs usually use a faster method of finding the maximum than computing all combinations, but may sometimes resort to this – particularly when the response surface is not 'smooth'.

H. The process can be viewed as a multi-parameter extension of the classic model of scientific measurement of a single parameter. In this the model would have one parameter (only) to be determined by measurement using data. Usually the model would be a simple formula. The optimisation model is congruent with the classic experimental model for this simple case – indeed, fitting a regression equation is exactly this process, with the regression parameters (a, b) of the model $y = ax + b$ being the values at the peak of the response surface. As the models become more complex (many lines of computer code) and the number of parameters becomes greater, this congruence disappears.

34. Optimisation of a Physically Based Watershed Model (2)

A. This would be an excellent response surface with a clear 'maximum'.

B. The values (k1*, k2*) would be the best in this two-parameter space.

C. True. In this area the model would be a poor fit to the observed data and varying k_1 and k_2 would not have much impact on how the model behaves.

D. The path of steepest ascent method would work well in this case and would be an efficient algorithm.

E. Usually if a response surface is drawn at a fine-enough scale there are roughness elements relating to rounding off, programming errors, underflow errors and overflow errors. Often some 'smoothing' is used to ease computation.

F. Optimisation becomes difficult when there are 'multiple peaks'. One approach is to use multiple starting points and see if one arrives at the same (k_1, k_2) pair. The 'peaks' reflect that there are a number of (k_1, k_2) pairs that give, effectively, the same fit.

G. The difficulty of finding the peak is that there may be multiple peaks or it may be 'flat terrain' with no peaks. You are, of course learning about your model and its parameter sensitivity (and sometimes you are learning more than you wanted to know). An automated system would usually have a number of starting points and see if they ended up on the same 'peak'. If they don't, one approach is to systematically map out the surface. Alternatively model alterations may be called for.

35. A Budyko-Hydrology World

A. A. Remember actual evaporation is about 0.6–0.7 of potential evaporation. Hence this makes the E/P ratio about 0.5.

B. D. Unless Ep/P has a value, all we know is that the evaporative demand exceeds precipitation, meaning that the system's evaporative dynamics are limited by the amount of water available to be evaporated. In heavy storms water infiltrates to depths below the rooting zone and this restricts the plant evaporation.

C. C. The ration of evaporation to precipitation leaves 25% of non-evaporated water free for streamflow. Ep/P is not relevant to this calculation if you already know E/P.

36. Sodic (and Other ic) Soils

A. Less common than sodic soils. Notwithstanding their magnesium-origins they would be generically referred to as 'sodic soils', and in the absence of chemical analysis the role of magnesium ions would not be recognised.

B. Such soils have a complex relationship with fresh water. Thus their clays may be stable in very pure water, unstable at low sodium or magnesium contents, and then stable at higher ionic contents. Often the presence of organic matter ('fulvic acids') stabilises them. In practical terms it is difficult to predict what will lead to instability, Earth moving that penetrates the organic layer is always dangerous because it allows water to pass beneath the organic layer and disperse the clays on an erosion face.

C. Fig. 5.36B shows soil piping. In this, fresh water flows through a preferential path (old root hole, etc.). The water disperses clay along the sides of the path, enlarging the hole. The tunnels get larger and larger until the soil above collapses.

D. Water from such areas may well be mildly brackish and discoloured by suspended clays.

E. Generally, Ca^{++} in the soil solution stabilises the clays. Thus, application of gypsum and building up of organic matter is the usual strategy. Occasionally mechanical barriers (erosion walls, etc.) are tried but the construction work associated with these may initiate new failures and often the foundations of the works tend to fail by undercutting.

F. Once gullying on sodic soil starts it will lead to a dendritic network of this type. The fact that you might have a model of its geometry is little compensation for the ugly gullying that is marring your watershed.

G. Because of the binding actions of deep roots and the high organic matter, 'erosion excursions' are less common in forested landscapes. However, they can occur. Reforestation is sometimes used to stabilise 'sodic landscapes' but trees tend to be intolerant of high sodium levels in soils. See also the next answer.

H. Sodic soils occur across the world and are usually found in low to medium rainfall areas. In higher rainfall areas the soluble ions have long been washed out. Thus, these are usually not an issue in high-rainfall areas (usually carrying native forest).

It is suggested as an exercise that students investigate the occurrence of sodic and other erosive soils in their locality.

37. Dryland Salinity

A. Mainly evapotranspiration that is reduced by the vegetation change. This is the primary hydrology change, however, there are changes to runoff generation processes, particularly overland flow.

B. Answer, B. Human intervention causes a natural state to change, hence it is a secondary process. Irrigation can be particularly associated with change.

C. Capillary rise under a tension gradient can bring saline water to the surface. Deep-rooted plants can be affected. If the rising water tables establish hydraulic connection with stream networks the streams may become saline.

D. Answer, C. While it may be dependent on the groundwater systems driving the salinity, planting in the recharge zones is aimed at diminishing the recharge of the groundwater system. Planting in the discharge zone is unlikely to work as this is where the salts are concentrated and will retard or prevent growth. Planting at high density all over the landscape is probably the best hydrologic option but is not usually compatible with agriculture.

E. Native vegetation was removed in both cases. The prairie grasses were removed to enable cropping. The tillage destroyed the soil structure and made the soil more easily erodible. Drought then caused crop failure and the bare, erodible soil literally blew away. Both examples have their origin in a fundamental misunderstanding of the landscape hydrology.

6 Essays and Projects

The foregoing five chapters have had a plethora of mainly technical and environmental material. This is valuable for the neophyte hydrologist, but it is only a part of the picture. Most hydrologic issues rather messily divide into technical, social, economic, and environmental components. The economic issues can, to a greater or lesser extent, be quantified but the social issues are usually matters of judgement. Exploration of these can be a valuable experience and addition to more formal studies. We have made some suggestions for programmes of this type and essay questions for those late-night study sessions. These days many of the issues (and places mentioned) can be studied on 'YouTube' and other videos too.

A Current Affairs Hydrology Scrapbook

This can be done as a personal exercise or class exercise. The task is to scan a newspaper/bulletin board and add articles that deal with some aspect of hydrology and watershed management. It is surprising how quickly the file builds up, and one becomes aware how pervasive such issues are. Hopefully it will make the things taught in class seem relevant.

A Few Suggested Topics for Assignments

Water is a universally needed good, and the richest, the poorest, the highest, and the lowest in society all need about the same amount of water to survive. Because of this and the omnipresence of rain, we have a vast number of structures to manage surface water. Some of these are thousands of years old (e.g. the aqueducts carrying water to Segovia in Spain, and 'tanks' in India).

© Leon Bren and Patrick Lane 2021. *Key Questions in Hydrology and Watershed Management* (L. Bren and P. Lane)
DOI: 10.1079/9781789249682.0006

This also means a vast and rich experience of hydrologic structures that have worked very well, not so well, or not at all. Our rivers, streams, and watersheds carry the footprints of these – just waiting for students to tease out what went right, what went wrong, and what are the lessons for the future in a great assignment. There are a few suggestions below.

Topic 1: The Lost Streams and Rivers

Consider Sydney, NSW or Hobart, Tasmania. The early European colonists rowed away from their sailing ships looking for a stream with 'sweet' (fresh) water. In the case of Sydney, they found 'the Tank Stream' and at Hobart the 'Hobart Rivulet'. In both cases they made a camp and then built huts, and before you know it a town had formed. Then, suddenly, the pretty little stream that they had valued so much was flooding their bark huts so, rather than relocating the huts, they made 'stream improvements' (aka drains) that they hoped would stop it flooding. This process continued apace until the stream was buried under pavement – giving rise to the concept of 'lost streams'. In some cases, the streams are small enough to be contained in 'drains'. In other cases (e.g. the Park River in Hartford, Connecticut) it is a major underground river and supports boat tours ('a unique experience, the darkness and the dripping, and the echoes...').

The task is to look at somewhere you know and map the streams that existed one or two centuries ago and see what has happened to them. You will find that many of the projects happened many generations ago, the people responsible are long-since deceased, and the organisations that manage these usually have little knowledge of them and the rationale for putting them underground. They may well be forgotten but they are still streams and, from time to time, will pop out and flood the people living along them.

It doesn't have to be an urban area either – many country streams and rivers have been deranged. In older parts 'canals' were used to transport goods and these tended to take water from one watershed to another. Miners often built tunnels across the neck of anabranches so that they could sluice gold out of the stream bed without having to deal with the usual flow. Watersheds have many old dams which pass water into 'races' or pipes which may still work but have been long forgotten about. The channels were often straightened as 'improvement'.

In some major cities much of the work has been done and there are often maps showing the lost stream network. Talk to the drainage managers in your town who can be amazingly helpful and will sometimes find old maps and construction plans. Once you have an idea of their location, walk along them and talk to older residents. And see what you can see of the 'lost streams'. You might consider what would need to be done to get the rivers 'found' and whether their riparian values could be recovered.

Topic 2: Flood Protection and Levees

The simplest method of avoiding flooding is to not live where it floods. But we can't all do that, so we resort to building dams, straightening streams 'to make them better flood-ways', and building earthen walls ('levees') to keep one side dry. Levees can do this but because they block off the flood plain, they make the flood on the other side of the levee much deeper than it would have been. And if it rains heavily then you have a problem of what to do with streamflow on the 'dry' side of the levee.

Your task would be to find an area prone to flooding, see what has been done to stop this, and document historic floods. In some cases (e.g. the Mississippi River) the levees can be huge and iconic structures; in others humble earthen walls protecting a few houses. But each has its hydrologic effect on the watersheds. You might look at which areas are being protected and which are being sacrificed and the relative long-term wealth of the communities. You might look at what happens if and when the levees ever fail. And you might examine the reason 'levee dynamiting' recurs as a practice in some parts of the world.

Topic 3: Dams – Their Rationale, Function, Failure, and Demolition

For watershed hydrology students, dams can provide a great and fascinating saga. The benefits and disbenefits of dams provide vast amounts of material. For some major dams, five decades after construction, the jury will still be out as to whether it has had a net benefit. Many dams have long passed their 'use-by date' but demolition is difficult and expensive and the question of rehabilitation of long-flooded areas is usually beyond everyone's experience.

The task would be to take a major river in some part of your home territory and to look at the dams on it – working dams, failed dams, abandoned dams, obsolete dams... As best you can, work out which dams meet some sort of human need and which ones don't, and then consider the way forward (including demolition and rehabilitation of the areas). Do they meet modern dam standards? Does the reason they were built still exist? What might be done to the previously flooded area if the dam is demolished? These are all key questions.

Topic 4: Hydrologic Disasters – What Went Wrong and Why?

We are all familiar with the stuff of the usual hydrologic disaster – houses floating down rivers, interiors of houses wrecked, cars piled on top of one another, the occasional loss of life, etc. Yet some analysts have argued that the real hydrologic disasters are economic or sociological in nature, and that the long-term costs of these far exceed the occasional property damage when the hydrologic system goes on a rampage.

There are many examples of human-induced changes in landscapes that have led to deleterious or disastrous impacts. Flooding related to land-use change (particularly deforestation), or flood impacts downstream due to river straightening or levees are some obvious examples. Others include the

'Dust Bowl' wind erosion, secondary salinity, and the myriad erosional outcomes from poor land-use practice. Students could pick one of these, analyse the physical processes behind these outcomes, and advance some alternative actions that would retain the land use, but produce a better hydrologic outcome. You might consider the socio-economic distribution of the costs (financial and otherwise) and benefits, too.

Topic 5: Balancing the Greater Hydrologic Good in Major Water Projects

Drought in Australia has led to many dams drying up and exposing sites that were once towns or homesteads. This has led to bitter memories from old residents about their forced relocation and the inadequate compensation paid. There are many examples around the world where, for the 'greater good of the country', verdant, productive areas were flooded. A recent thesis on the formation of Quabbin Dam to supply Boston (USA) concluded that many of the displaced residents were surprisingly objective about this and were prepared to make sacrifices in their own life for the 'greater good'.

Your task, as students, is to look at some of these past epics in your own locality and see what can be learnt for the future. Is financial compensation an adequate recompense? Perhaps there is no answer? The students' task is to tease out the issues, decide where the national interest lay (or lies), and to work out what could or should be done in the future.

Topic 6: Empathetic vs Cold-hearted Hydrology

The flood has receded and, for the first time, people can get in. The two hydrologists look at the pathetic house wreckage. A mud-stained, sodden cloth doll of an unknown little girl lies in the midst of debris from a collapsed wall surrounding an overturned stove. Hydrologist One is moved to tears. 'I wonder how the little girl fared – such a human tragedy', they muse. Hydrologist Two looks with disgust at the wreckage. 'Weeks of cleaning up here; why do people insist on living in such flood-prone places? They could of at least got the little girl a plastic doll so we could dry it more easily' is their muse. So, here we have the contrast of the warm-hearted approach (perhaps people-centric, emotional) and the cold-hearted approach (technocratic, outcome-focused). In hydrology there is plenty of room for both approaches and both can be effective in the long term; it can be argued that, in the long term, the second approach can remove much of the need for the first approach.

The task is to look at local or international hydrologic disasters (floods, landslides, etc.) and to analyse them in a warm-hearted, emotional way and a dispassionate way. Are the victims to blame for their own fate, or are they powerless people in a larger theatre? You might also consider how the clever hydrologist might learn to walk a tight line between these poles of analytical behaviour.

210

Topic 7: The World's 'Water Towers'

Classically, a water tower is an elevated building supporting a water tank. This pressurises a water distribution system. They are a common feature in country towns in flat landscapes around the world. More recently the concept of the world's 'water towers' has appeared in hydrology. The argument is that mountains are the 'water towers' of the world, supplying a substantial part of both natural and anthropogenic water demands. They are highly sensitive and prone to climate change.

In particular, the 'Asian Water Towers' are considered of great importance because of the huge population dependent on them and their sensitivity to climate change. Thus the 'Indus Water Tower' is said to be the most relied-upon glacier-based water system in the world. This leads to the possibility of climate change creating large numbers of 'water refugees'. Some hydrologists argue that 'water towers' is dressing up old concepts in new clothing or, at best, a rather 'strained analogy'. Others argue that it is a new and coherent way of looking at the world and coming to grips with the realities of climate change.

As students, your task would be to examine the concept, to see if you can identify 'water towers' in your home country and elucidate likely threats to them in the near and far future.

Topic 8: Climate Change Impacts

The task is to see what climate change may do to the system of hydrology. This might entail gathering data on the system hydrologic properties, vegetation/land use, etc., then seeking out climate projections, and then applying those projections to the hydrologic functioning of the system of interest. This could include fundamental processes of evapotranspiration, streamflow, and water quality, land-use changes that may be possible or required, and community socio-economic impacts.

Another option might be to give a climate change scenario, and then pose a problem to the students as if they were professional hydrologists. One group may have $50,000 and 3 months to do the job, another group might have $250,000 and 1 year to do the job, another group $5,000,000 and 5 years to do the job. Students can think about approaching a project with layers of resourcing and time and what can be achieved with the resources, what differences in approach can be taken, and what certainty of outcomes might be reached.